INSCAPE AND LANDSCAPE

INSCAPE AND LANDSCAPE

The Human Perception of Environment

Pierre Dansereau

Columbia University Press
New York and London 1975

Adapted from six radio programs broadcast in the Fall of 1972 by the Canadian Broadcasting Corporation as the twelfth series of Massey Lectures. These lectures, named in honor of the Rt. Hon. Vincent Massey, former Governor General of Canada, were begun by CBC in 1961 to enable distinguished authorities in fields of general importance to present the results of original study or research.

Pierre Dansereau is Professor of Ecology at the University of Quebec at Montreal.

Published in Canada in 1973 by the Canadian Broadcasting Corporation
Columbia University Press edition 1975

Library of Congress Cataloging in Publication Data

Dansereau, Pierre Mackay, 1911–
    Inscape and landscape.

    Originally published by the Canadian Broadcasting
Company, Toronto, in series: Massey lectures, 12th ser.
    Bibliography: p. 106
        1. Human ecology—Addresses, essays, lectures.
    2. Geographical perception—Addresses, essays, lectures.
    3. Ecology—Addresses, essays, lectures. I. Title.
    II. Series: Massey lectures; 12th ser.
    GF47.D35 1975   301.31   75-9990
    ISBN 0-231-03991-3
    ISBN 0-231-03992-1 pbk.

INTRODUCTION

The world of man is expanding. This generation is living the fulfillment of an early dream of mankind: it has set foot on the moon. We are still in the grasp of the adolescent stupor that accompanies the discovery of our own power. And yet, wisdom is of the essence if we are to redress the course that now leads the human species to suicide. This can only be done if we develop a full consciousness of our growing influence over our environment, accept new rules of stewardship, and develop a responsible plan of environmental management.

The science of ecology has much to contribute to this outlook if it can lend itself to a bold integration of environmental data and interpretation thereof and if it can gather the human impact within its scope. This is both an indispensable preliminary and an ongoing requisite for the re-setting of our common goals.

Modern science is in possession of considerable knowledge concerning the material realities of almost the whole Earth. Analyses of landscape are numerous. We have sounded ocean masses, calibrated their salinities and temperatures, mapped their currents; we have charted the rivers, lakes, waterfalls, and ponds. We know the distribution of mountains, valleys, and plains and of the forests, grasslands, and deserts with their characteristic vegetation and faunas. But we are not fully aware that land-uses have variously modified the natural balances and have created new ones.

In words and in pictures, in charts and in maps, we have given ourselves a large repertory of world landscapes. Geologists, engineers, architects, biologists, economists, and others have minutely analyzed the processes whereby man has cast his stamp upon these vistas and have passed judgment on their productivity, their stability, their beauty.

The environmental disorder which we face at this time is the result of the discrepancy between man's intentions and his achievements. The loss of balance goes back at least as far as the industrial revolution. In fact, there is evidence of it in earlier cycles of civilization, such as the Mesopotamian and the Mayan. It is all the more humiliating that the "environmental scare" did not really hit the industrialized countries at the level of the general public until the 1960s, and is not likely to strike very hard at the consciousness of the "developing" countries for some time to come.

The comprehension of ecological harmony and the apprehension of ecological disaster is a matter of bringing together the accumulated knowledge of scientists and projecting it into the inner reflection of the individual and of the people.

My concern, therefore, is as much with the inscape as it is with the landscape, as much with the human perception of environment as with man's impact on nature. Indeed I view the inscape/landscape process as a cycle. Man, from Magdalenian to modern times, has had a selective perception of the world about him and in turn a highly discriminating way of modelling the landscape to match his inner vision.

"Inscape" may be an unfamiliar word in this context. It was coined by a poet, not an ecologist or a geographer. Gerard Manley Hopkins recorded his contemplations of nature in diaries, letters, poems, drawings, and even in music. This filtering inward from nature to man, upward from the subconscious to the conscious, and from perception to design and implementation, is indeed what happens to the agriculturist, the forester, the engineer, the town planner. The pathway of sensorial impression to material interference is strewn with an imagery that makes the inscape a template for the reshaping of the landscape.

I therefore propose to focus attention on the successive landmarks of this itinerary by examining the operations involved in: perception, training, research, power, planning, and management. I want to do so with constant reference to the real world around us, sometimes turning to the virgin tundras or forests, sometimes to the lowing pastoral grass-lands, sometimes to the clamouring city. Heartened by our spectacular and calculated successes, and alarmed by our mindless failures, we can only question the values that have guided our human societies. A rational appraisal of the remaining resources is very sobering indeed and the spectacle of our wasteful ways calls for a drastic change in our common goals. I shall argue in favour of a self-imposed limitation of human population and a reconsideration of economic growth and express the hope that the quality of life can be maintained through a joyous austerity.

P. D.

CONTENTS

The consciousness of self, the knowledge of life, the assurance of death are raised, in the human species, to the highest pitch the animal kingdom has known so far. It is also fair to surmise that kinship with the dead and invocation of a higher and invisible power, not to mention hope of an after-life, bind the members of the human species together in a unique fashion (Dobzhansky 1967).

Such emotional stresses have the strongest influence on what man does to himself and on the way in which he exploits and manages his environment.

The theme of impending death looms very large in the Spanish and the Japanese cultures; the theme of sweetly fleeting time pervades Polynesia; the theme of physical prowess inspires the Germans and the Masaï. Each culture has some privileged focus that adjusts its vision to its material environment. The bullfight, the lion hunt, the Wagnerian opera, the painted screen, the hula dance, each has only one authentic décor.

As we search in the history of our species for the images that man has given of himself, we experience a certain recognition in the behaviour of the reindeer and bison hunters who have graven their silhouettes on the walls of Lascaux and Altamira. We enter into the mass movements of these frescoes where man and beast are joined in a ballet of war and festivity, of love and sacrifice. The anthropologists are unravelling the messages that these remote ancestors have sent down to us and which are echoed through the ages in forms ever closer to our own language. Thus Leroi-Gourhan (1965) offers a grammar of sexual symbols that prepares us to understand the complex motivations of so-called primitives who have met some of the greatest tests of adaptation to environment by producing symbolic rather than material responses to it. If it is true that the domestication of animals, as Hahn (1896) and Isaac (1962) have suggested, is of religious rather than economic or utilitarian origin, must we not give considerable weight in all circumstances to the interior imagery?

The Greeks, although they valued reason above all and submitted natural phenomena to rigorous examination, never ceased to play the flutes of Pan in the background. For them no explanation of the place and role of man was convincing without the exemplary mysteries of heroes and gods, and it was the dionysiac game which best expressed their communion with nature.

Christianity, by spreading the mystique of the celestial home and of personal salvation, drew man away from nature, made him relate his everyday gestures to an assumption that took him out of his environment. Whereas the agapes of the first Christians had a lingering dionysiac undertone, the discipline of the Middle Ages recentered their unanimity in formal obedience to precept and in the supremacy of the supernatural. This opened the door to contempt for physiological needs and it made Christian societies restrict their accords with the physical world.

Even if the Renaissance rehabilitated direct observation and the Reformation revalidated individual conscience, they did not very sensibly modify the methodological framework of the perception of environment. Christian celebration continued to be an act of sublimation.

Modern science, which developed mostly in occidental Judeo-Christian countries, still placed man in the centre of nature unless it be at the summit, and resolutely rationalized what separated him from the plants ("Man is a thinking reed", said Pascal), and the animals ("a reasonable animal", again according to Pascal), and according to Lamartine:

Borné dans sa nature, infini dans ses voeux,
L'homme est un dieu tombé qui se souvient des cieux.

Emerson (1836) and the transcendentalists, in spite of their diffidence towards the puritanical ethic, did not sever themselves from it very strongly. "The greatest poverty is not to live in a physical world", said Emerson. But he added, concerning the contemplation of landscape, "It is necessary to use these pleasures with great temperance." He also said: "Nature always wears the colours of the spirit." (This approaches Hopkins.) And: "There is a kind of contempt of the landscape felt by him who has just lost by death a dear friend. The sky is less grand as it shuts down over less worth in the population." (Lamartine: "Un seul être vous manque et tout est dépeuplé.")

It is not astonishing that Darwin (1859) rose as a menace to this marriage of rationalism and romanticism which had invested our species with such singular privileges. Thomas H. Huxley's first book (1863) was entitled *Evidence as to Man's Place in Nature*, and it seemed to undermine the very foundations of the churches and of occidental societies. One must add that if that work finds echoes today, for instance with Sears (1957), Bates (1964), Darling (1964), Glacken (1966), its point of view has not generally imposed itself, far from it. In other words, the supremacy of man has retained our attention much more than his integration into the environment.

In the first half of the present century, purely scientific curiosity (which turns out to be so urgent for our very survival) did not exert itself fully in the study of conflict resolutions between the hereditary potential of man and the stresses

3

of his environment. The warnings of Malthus (1798) were re-issued from time to time, but they seemed to derive from a certain moralism rather than to be the voice of science itself. In truth, the advocates of conservatism were almost all naturalists or else were inspired by them (Teddy Roosevelt for example) and they invoked the evangelical principle of stewardship. They often launched an appeal to the responsibility of man towards animals and plants and eventually toward the soil itself and the water and air which nourish us all. Maybe naturists and vegetarians should be mentioned in this connection. Their sub-philosophy, however, is hardly taken seriously even when it borrows illustrious voices like that of George Bernard Shaw who responded to popular masochism with his own form of humanitarian contempt.

As we look back upon the unfolding of man's contacts with his environment, we must ask ourselves many questions concerning the quality and selectivity of his perceptions on the one hand and on the ever-increasing strength of his power on the other.

The relation of perception to survival comes first. Thus it is no wonder that Eskimos have a number of words to designate the physical states of snow and ice (that is useful to the winter-sportsmen), that the Polynesians invented a vocabulary (now adopted by the geologists) to differentiate the textures of cooling lava flows. The Maoris and the Brazilian Indians have names for quite inconspicuous plants, and the Central African hunters recognize all the birds and the beasts on sight.

The many ways in which this knowledge is achieved involves the alertness of all senses, not only to the shape, form, and colour of foliage or plumage, but to the feel and taste of the bark, the leaf, the fruit, and the flesh. The resilience of the fibre, the brittleness of the bone, all serve to identify the "species", to trigger instant communication between man and plant, man and animal. And, of course, between man and rock or soil. One has only to gaze intently at the rough surface of the caves of Lascaux to marvel at the knowledge the artists had of the grain of the stone wall.

If we turn to socialized units and to modern man himself we can better seize the nature of this apprehension of rock, plant, and animal in its more explicit form. The essentially non-utilitarian grasp of scientific identification reaches its full dimensions in the naturalists' description. The great explorers of the eighteenth and nineteenth centuries convey at once the accurateness of their perception and the joy of their recognition of things-as-they-are. Humboldt, DeCandolle, Darwin, and Audubon are totally involved in their descriptions, in acute mental concentration to be sure, but they are visibly borne by an emotional surge that denies none of the senses. To be sure, problems always lurk in the background and they are alternately in soft and in sharp focus. A previous question was posed about this bedrock, this lichen, this bird, and it will recur again and again, but it will also be shaded out by the joy of apprehending texture, shape, colour, and movement in their reality and in their beauty.

Thus the consciousness of the sophisticated naturalist, casually touching a glaciated rock, a wet leaf, a furry pelt, superimposes itself on that of the cave-dweller, the farmer, the hunter who had already appropriated the land. The scientist's codifying of sensorial knowledge reaches a higher pitch of retainable experience; it can remain as personal as that of the wild man while acquiring a better state of communicability.

The apprehension of form and structure and the images of specific repetition are often triggered off by sensorial shock or gratification and this is why perception is so closely linked first to survival, then to utilization. The resistance and yield of the substratum, the texture and taste of the plants, the fleetness and availability of animals are all imprinted upon the mind of man struggling for existence, and they are at the origin of his devices for exploitation. It is all too easy to conclude that post-neolithic men have lost their sense of sight, touch, smell, and taste, since they have merely exploited previous discoveries and have not experimented on their own. Is it not rather shocking to reflect that all staple plants (rice, wheat, potatoes, maize, sugarcane, etc.) were domesticated in neolithic times and that modern science has discovered no new wilding capable of providing a major food resource? Instead, "science", as we know it, has merely bred bigger and better, and hardier and more productive strains.

It could well be that Europeans who have settled in the agriculturally more fertile lands of Africa, America, Asia, and Australasia suffer from extreme reduction of consciousness towards rocks, plants, and animals. Their knowledge is often confined to the relatively small stock of cultivable plants and domesticated animals they brought with them to North America or New Zealand and the almost equally small number of agricultural plants and animals they exploited in the tropics. It is not accompanied by the persistent and rich folklore of the European lands where modern agriculture originated. On other continents the repertory of French, English, Dutch, Portuguese, and Spanish names for wild plants (native and introduced) is rather short and, for the most part, recast from European analogies. The latter, of course, are of the highest interest to botanists, and it is most often they and not the farmers and hunters who have provided vernacular names, some of which have found their way into popular usage. Shield-fern, heart's-ease, touch-me-not, forget-me-not, bluebell, hornwort, are English categories, as mouron, amarante, benoîte, épilobe are French. Some of the more noticeable, useful, or noxious plants have received new names, for instance, in French Canada: pruche, épinette, bouquet-de-la-veuve, vignette, jargeau, bleuet, vinaigrier, herbe-à-la-puce, sang-dragon, thé-du-Labrador, cenellier, herbe-à-poux, graquias; and in Anglo-America: leatherleaf, snowberry, lambkill, Indian-pipe, hay-scented fern, pitcher-plant, squirrel corn, blood-root, pin-weed. And fortunately some Indian names have fallen into common usage: pimbina, maskouabina, atoca, pipsissewa, saskatoon, cohosh, tamarack.

Such is perception at the first level, that of sensorial apprehension which results in the identification of natural objects, individually focused upon and perceived.

At one remove is the selective utilization of rock, plant, and animal by man. How do the races of man, inhabiting different landscapes, exploit water, soil, vegetation, and animal life? How much do they know, what do they use, and how much on the one hand for bodily necessities and on the other to fulfill psychological needs?

When such a question is posed by an ecologist rather than by an anthropologist, he is likely to look first at the environment and second at man. Scanning the landscape he wonders what its resources are: whether varied, abundant, renewable, accessible. Let us consider three series of examples where the land is characterized principally by its minerals, by its vegetation, by its animal life.

The great inland deserts are all sky and rock and sand. The infrequent showers are followed by rapid evaporation and by infiltration into widely scattered and fluctuating pools and ephemeral streams. On the edge of these are borne fringes of greenery, whereas elsewhere bristling or spiny shrubs reflect the underground network of available moisture. Outward, towards the steppe, hard-leaved silvery grasses appear, and where winter rainfalls increase, thornbushes or broadleaved scrubs and savanas; the small rodents, blond as the sand or rufous as the rock, nestle in the clumps of grass, burrow in the crevices and feed the swooping rapacious birds. The oases fluctuate in shades of green, around the glaucous eye of their ponds. Palms or wattles shade the ring of grass. It is a mineral landscape under the full impact of scorching sun at midday and chilly nights condensing dew on the ground and on the blades of the leaves. The stones are cut to sharp edges by the blowing sand. Under extreme heat/water ratios the upwelling capillary streamlets deposit a crust of salt on the surface. In the milder climes, the tight mounds of dwarf shrub or grass hold lumps of decomposing plants and animal excreta that

7

nourish the plants that feed the small, burrowed herbivores.

In this sparsely populated world, man is a wanderer among wind, sand, and stars. The "terre des hommes" of St. Exupéry starts here, in this elemental contact with the roughing sand, the stunning heat, the sheltering firmament. Man dreams of water, of fresh greenery, of succulent fruit, of cool milk and roasted meat. The hard discipline of scarcity governs his every move. The camel is his ally, for travel, for wool, and for sheer companionship. And so, in the milder periphery, are the sheep, the goats, the donkey. Nomadic life revolves around protection from heat and cold by woven wool, care and service of beasts to be fed and watered, by moving constantly. Stone shelters and tents protect the human horde against the elements, allow storage of cheese and meat, of precious fuel, and facilitate the long labours of weaving, of basketry, of tanning and leatherwork. The open sweep of the land, the mirage of the oasis, the darting presence of small rodents, the infrequent soaring of a bird, combine with the reassuring presence of the beasts of burden. In the quiet of the shelter, the artisan recalls these shapes and draws them on the walls, in the sand, on the hilt of his instruments, on the surface of his pottery, in the weave of his carpets and drapes and clothing. That is unless a religious constraint forbids graven images, and then the shapes of plants, animals, and man are left out or reduced to symbolic allusions. Or can it possibly be argued that the prevalence of the mineral world over the living sphere is conducive to the abstract design that the Arab cultures have raised to such a high level of sophistication? This is an attractive idea, in view of the fact that, for instance, the desert Navajos have successfully surrounded themselves with useful and non-utilitarian objects incorporating both representative and abstract forms.

At another extreme in the mineral landscape, the icy wastes of the Arctic have elicited almost opposite responses. There is no scarcity of water, but of fuel. The stress of low temperatures has been met at the genetic level by plants, beasts, and man. Adaptations to extreme cold are quite visible in the size and branching of plants, in the ratio of volume to surface exposed, and in the metabolism of fatty tissue in animals and man. Crustose lichens, dwarf heaths, ptarmigans, polar bears, and Eskimos are descended from many generations stressed to the rigour of the long and barren winter night and the long summer day with its teeming abundance. As with the creatures of the desert, environmental adversity is met in many different ways, from over-feeding to accumulation of reserves (within the body or in caches), from hibernation to migration.

Eskimos and Laplanders are highly conscious of natural objects. Their snow-vocabulary is the richest in the world. If their botanical folklore is less impressive, this may reflect the bypassing of plant-food in their almost entirely carnivorous diet. Their keen assessment of the habits of animals has something fraternal about it. Indeed their willingness to learn courage and skill from their prey is reflected in their hunting habits and in their domestic customs. They hang the pelt of the arctic weasel around the neck of their children, for they have observed this tiniest of carnivores to be the most able hunter. From the cosmic aurora borealis to the creeping cranberry and the flying duck, their landscape is full of moving objects that signify. The seal and the caribou are, of course, privileged; the Eskimo have learned to process all parts: bony structure, skin, viscera, muscle. The mixture of awe, respect, and of a certain tenderness emerges in the new-found techniques of the sculptors and engravers of the Canadian Arctic.

Turning to a second panel of landscapes where it is the vegetal rather than the mineral processes that meet the eye, we face the huge biomass of the tropical rainforest, the "green hell" where the constraints can well be said to equal those of the Arctic or the desert. The penetration of heat into the very depths of the soil has altered bedrocks to a greater extent than anywhere else. The drenching rains and the great network of rivers supply a flow of moisture that ascends through the living mass and hangs above the steaming forests. The vegetation of the Amazon and the Congo basins is the richest in the world, but its very diversity causes a confused scattering of energy when specialized animals, such as man, must fulfill their needs by gathering. Colin Turnbull, in his intimate study of the Central African Pygmies, makes us understand the habits and the psychology of forest people. Their verbal repertory may be limited, but their knowledge is extensive when it comes to fungi, flowers, fruits, herbs, trees on which they feed, with which they build their shelters, mould their few implements, and design their scanty clothing. The forest, and nothing but the forest, is their world, as witness the panic of a young man on a trip through the grassy savana! He feels that no human can live in open spaces of this kind, without the protection of trees!

On the contrary, the Plains Indians who had the freedom of the great grasslands of north-central North America avoided the forest. The deep black soil nourished the lush tall grass of the Prairie; very green in the spring, turning to brown in the late summer, and mixed with goldenrod and other bright-coloured forbs, it abundantly fed the roaming herds of buffalo. The hide, the meat, the horns and hooves of this beast, and of the deer and the antelope, provided food, shelter, and ornament. Marsh herbs yielded starch and sweet berries. The devastating graze of the buffaloes and periodic fires rejuvenated the vegetation and renewed the stock of plants, opened up new habitats to migrant birds. The hunter's calendar, in this rich, rolling landscape, was very full and quite varied. The sea of grass had many seasonal harvests. It comes as something of a shock that European settlers first perceived it as a "desert". In their cultural perspective, fertile land had to be reclaimed from the forest, and it is on the wooded islands of the Prairie's edge that they first sowed their wheat and barley (Shelford 1944).

In a third panel, we would seek man among the animals, in a landscape where birds or mammals represent a comparatively large mass. Just for the sake of horror, we may cast a glance at the Pribilof Islands where the heavy-breathing masses of sea-lions literally cover the rocky shores, plunging into the battering surf and diving for food, surging back mightily on the polished promontories only to be assaulted by small groups of men with heavy wooden bats. A similar ecological phenomenon can be witnessed in late winter in the Gulf of St. Lawrence where snow-white baby seals adrift on the floe-ice are likewise killed in the interests of the fur industry. I need not pause at this stage to question the ethics or even the ecological wisdom of these practices. I intend to do so later on. But I find it hard to rejoin my purpose of sounding the perceptual dimensions of man in this animal-dominated environment without paralleling the promptings of conscience with those of consciousness. I doubt whether the psychology of the Pribilof hunters and of those of the Gulf of St. Lawrence can be assimilated to that of the whalers.

Having lived a short while with the sweet-tempered, courageous, and patient Azoreans who still hunt the whale in their long and narrow canoes, I recall the ritualistic obeisance of their response to the signals of the weather men and of the watchers in their towers. Scanning the sea, pursuing their quarry, flexing their muscles for the harpoon throw, dragging in the wake of the mighty beast, and leading its monstrous bulk to the harbour is heroic work. Above all it is teamwork of a most exceptional kind, discipline of a quality unknown in our research laboratories. Many of them die and some are crippled. But far from them the blasphemous resentment of Captain Ahab! The peaceful operation of separating bone from blubber, skin, and oil goes on for days and the segregated masses of precious animal material are processed and packed for use as ivory, oil, ambergris, fertilizer, and even meat.

Turning to another animal-dominated landscape, we find a great variety in the high plains of Central Africa. The flat or gently rolling land is cut by winding rivers often bordered by a row of trees, a galleria forest, that may extend to the edge of lowland reedy marshes, whereas the upland is clothed with grass (alternately green and brown with the fluctuating rainfall), and dotted with rather evenly spaced small, spreading trees, such as acacias, or islands of trees and shrubs. It is in these groves that the great predators dwell, the lion and his family. Out from the den onto the grassy savana, the hunting is very rich: water buffalo, zebra, and many species of antelope, all of them gregarious and hungrily moving from clump to clump of grass, suddenly triggered into co-ordinated scurry and flight. This great orchestra of the hunt is accompanied by the basso of the oblivious, lumbering elephants, the clanging splash and wallow of the hippopotamus, the sibilant darting of the wild pigs and the rodents, the prowling of the hyena, the chipping of small birds, and the heavy flapping of the vultures, the strident hum of the insects, the clicking of grasshoppers. Animal life at all levels burrows, creeps, climbs, flits, flies, runs, swims, all the while exploiting water, soil, roots, buds, bark, leaves, and fruit and of course other animals alive and dead. Man-the-hunter inserts himself into this clockwork of mineral–vegetal–animal transformations of matter, he has to know what stings, what bites, what poisons, what moves. He has to know his place in the competing order for food. Well-adjusted people like the Masaï have a rhythm that fits them well and permanently into the biocoenosis. Like the lion and other predators they have their order of priority in food requirements: whereas the lion prefers zebra to antelope, it is the other way around with man. In fact, the Masaï have other needs as well: their fondness for adornment which leads them to paint their bodies and dye their hair with red earth is accompanied by a need for proving their virility by killing a lion. Joseph Kessel has told this tale very dramatically, and spelled out the conflict of ancient ritual and modern conservation.

Thus, in various landscapes, where the mosaic of mineral–vegetal–animal elements is so variously distributed, the repertory of resources that can be directly or indirectly manipulated by man is very different. In the first place, the predominance of any given element (moving sand, lush vegetation, teeming animal life) sharpens the perception by an unrelenting sensorial impact. This is in itself a constraint. It is an invitation to acceptance or rejection. It is something to be used or to be overcome. It is something, above all, that calls for management.

The examples I have given, in order to emphasize man's perception of nature in three main types of environment, were deliberately chosen in times and places where the landscape as a whole had not been re-shaped by man to any considerable extent. These three panels will be scanned anew in order to consider the questions that are to follow, in subsequent chapters.

The first one, posed above, concerns man's attitude. The second will review the way in which these perceptions have been codified and can variously serve the purposes of education and training. The third will probe our understanding of nature and of ourselves within it by means of research. The fourth will run the gamut of processes involved in man's growing power in shaping the environment. The fifth will concern itself with the consequences of socialization and the emergence of design and planning. The sixth will review a variety of inscape/landscape patterns so as to consider where we stand in the application of our knowledge to the art of environmental management.

The discovery of the material world by man is an unfinished business. At this late hour, we do not have a complete inventory of the rocks, plants, and animals that occupy our planet. Even less have we fitted the distribution of species into the niches of environment, for we do not really know how many kinds of environment there are. This is a circular process, of course, since we must know the nature of the pieces (rock, plant, animal) and something of their dynamics before we can assess the outstanding characteristics of whole units. But we must, on the other hand, make some provisional statement concerning the spatial units too.

The variety of ways in which we learn increases and diversifies as the treasury of mankind accumulates. Indeed our reflections on the learning process itself are a part of the aggregation. I cannot well explore the whole of this field or even try to summarize the findings of pyschologists and philosophers, of epistemologists and semanticists, but I will try to relate my own intuitions on the quality of perception as it applies to environment as a whole and to the units that compose it.

I have already attempted to score a few points, in the previous chapter, by alluding to the human response within several different ecological cycles. I will now attempt to be more systematic and to review the tasks of description and interpretation, the tests of functional groups, and the means of storage of information. Families, societies, schools, states, and other institutions are the depositories of all this cultural wealth; and they also contain the centres of decision where differential exploitation is planned.

It may be well to re-emphasize that the fabric of this discourse on inscape and landscape consists in a warp that is provided by the naturalist in his attempt to tighten the thread of objective reality conveyed through science, and to draw across it the woof of human perceptions and implementations. The warp is the science of ecology, becoming a more broadly based environmental science, the woof is a socio-economic discipline that may soon deserve to call itself human ecology.

In order, therefore, to draw the threads in a consistent manner, and to firm up the tissue, I shall follow the pathways of human endeavour in taxonomy, ecology, and chorology (the science of spatial distribution). Across these scientific achievements I shall attempt to draw the cross-references of anthropology, sociology, geography, pedagogy, ethology, and psychology.

First comes the naming habit. Taxonomy is of the essence. Those of us who are both born and trained naturalists have a dream of the privileged first man who encountered stones and flowers and butterflies never seen before and to each gave a name. And even in this generation professional taxonomists are completing the task and finding unknown fungi, flowering plants, and insects.

But, naming is describing. It does not suffice to define the attributes that distinguish one object from others that look like it and from those that do not. A white rose is different from a red rose, and yet both are roses. And so a yellow and a blue violet, a black and a gray squirrel, a white-headed and a golden eagle, a white and a black man. But inevitably, taxonomy is a search for order. And on what may natural order be based? What true functions are revealed by form? Is there indeed a relationship between form and function?

I was trained in one of the world's best schools of taxonomy and my teacher, who was a solipsist (and who wrote very cogently on the subject), did not believe that classifications reflected anything beyond the convenience of the user. Unwilling to give up the quest for detection of evolutionary processes and to make them visible in schemes of classification, I was not able to accept this view. In taxonomy, as in so many other disciplines (not excluding philosophy), it is quite possibly the striving rather than the achievement that counts most for many of us. Whereas we cannot hope to detect all of the "missing links", it is fair to plot the relative positions of individual kinds of rock, plant, and animal (and man) in a hypothetical network of mutual derivation.

Actually the immediate need for tagging and the eventual requirement for classifying call for distinct operations. It is good enough to key out different species of cherry or field-mice on the basis of size of fruit or shape of ears, whereas we know full well that such features are relatively unimportant in the evolutionary sequences and that more significant steps were taken in the course of descent which are reflected in less visible characters. But it is more important to the viewer in a given landscape to distinguish between the species than to trace their genealogy. The latter cannot be perceived directly, although the trained geneticist will sense it. But our present concern is definitely more ecological than evolutionary, even if the genetic dimension, for some privileged observers, provides interesting clues.

In many parts of the world, especially North America and Europe, the inventories of rock, plant, and animal species is very advanced, and in some (the British Isles, Eastern North America) so nearly complete as to have yielded almost no novelties in the many recent years of exploration. Such knowledge, garnered by scientists, is available in the form of collections carefully and authoritatively labelled in museums, and in various compendia listing the known presence, in a given area, of the various kinds of rock, plant, or animal. "Floras" (lists and descriptions of plants) and "faunas" (lists and descriptions of animals) are on the desk of all practising naturalists and are also accessible to other users, such as foresters, agriculturists, geographers, physicians, pharmacists, who have mastered the technical language.

By and large, these catalogues contain morphological descriptions that living and preserved specimens are expected to match, and which can be verified with the naked eye or under low magnification, thereby enabling identification. This will allow the landscape analyst to establish the exact amount of diversity in his area and to know what the dominant species are.

The information contained in a flora, however, usually goes well beyond such a narrow purpose. In the first place, the specific units are ordained according to an accepted taxonomic system, itself based upon a phylogenetic scheme that recognizes many orders of magnitude. In plants this will involve variety, species, genus, family, order, and class. The higher units (class, order) represent major steps in evolution, and they can be dated as to their approximate time of emergence, whereas the lowermost (species, variety) can be presumed to be more variable and more recent, almost surely more localized. It does not follow, however, that specific members of the older classes are themselves older than species of the younger ones. Thus, many Algae, Ferns, and Conifers may be of as recent origin as some of the Flowering Plants with which they are growing at this time.

However, the relative weight of Algae, Fungi, Ferns, etc., in a given landscape is of interest in its physiological implications. The tapping mechanisms of plants without chlorophyll or without flowers, their growth-form and growth rate, their competing powers, their means of resisting cold or drought are all inscribed in the status of their class. It is through this bias that perception will change gear, from knowledge of identity gained through instruction to direct observation in the field. This too is bound to be a circular process since it evokes the great theme of form and function, and thereby invites comparison of the evolutionary pathways that have led to similar morpho-physiological developments in a singular selective environment. Thus an African and an American desert, an Australasian and a European alpine barren will be populated by plants that exhibit similar branching and leafing habits and other exploitive devices, although they are frequently quite unrelated genetically. These forms (often implying an inkling of their function) are readily perceived even by the casual observer. The candelabra cacti of America (spurges in Africa), the spiny shrubs, the brightly coloured "belly-plants" (ephemerals that follow an infrequent rain) all obviously respond to heat and drought. The twiggy, small-leaved, trailing woody plants, the compact cushion-plants, the tight rosettes with their vividly coloured flowers, the rock-encrusting lichens, are no less readily identified with the alpine habitat. And as for palms, and lacy-crowned tree-ferns. . . .

I could easily evoke quite a few other botanical types and relate them to a spontaneous triggering of environmental recognition in everyman's inner imagery. This reflex is no doubt very involved, and it is hard to say to what extent it is determined by actual scientific knowledge, how much by other forms of experience, personal or vicarious. We are, all of us, profoundly stamped by visual and other sensorial associations. I can testify, for instance, that many years of botanical study and of plant-ecological research have not marred my preferred vision of the forest as shown in the Gustave Doré engravings that decorated my childhood albums.

These examples taken from the plant kingdom are parallel to many in the mineral and animal worlds.

Such elements of the mineral fabric as water, gold, iron, granite, sand, limestone; such land-forms as cliffs, dunes, estuaries, plains, mountains have some generally shared resonance, but even more of a subjective meaning closely tied to individual or collective experience. The Swiss and the Dutch simply do not apprehend the flatland and the mountain in the same way; the Hawaiians, the Neapolitans, the Azoreans, and the Congolese do not look at volcanoes in the same light. The Sherpas and the Eskimos have a different attitude towards ice.

The animals, of course, lend themselves not only to perception of presence but also to impersonation and allegory. Naturalists and biologists have done their best to verify and more often to disprove the mythical character of the cat, the wolf, the skunk, the deer, the peacock, etc. Fabulists, storytellers, and comicstrip artists have generally responded to existing prejudices. Neither Aesop's nor La Fontaine's, nor Orwell's, zoology is very sound, nor need it be. Thornton W. Burgess' ecology is fairly good, but Kipling's is much better integrated. As for Walt Disney and Walt Kelly's cartoons, if they show little insight into the lives of animals, they flatter the prejudice of the average reader by stereotyping a noisy duck, a nosey mouse, a naively inquisitive opossum, a preening skunk, a noble horse, a faithful dog, a wise owl, etc. There is little doubt that those who sponsor well-considered wildlife legislation, based on reliable, objective research, have to contend bitterly at times with the myth of the big bad wolf, the deceitful fox, the treacherous hyena, the sinister crow, and other puppets of national or international folklore.

The knowledge of naturalists is there, in the museums, in the books, on the screen, in the schools, but it is not always made to signify. Those who have the knowledge at first hand have not also been capable of conveying the message that could have saved the passenger pigeon and the great auk; that could have prevented the senseless slaughter of elephants, bears, wolves, coyotes; that might have prevented the poisoning of the condors and eagles; that could counter the rapacious power of the fur trade.

It is worth noting that these storehouses of taxonomic knowledge are menaced, at this time, for lack of support and personnel. I shall return to this topic in the next chapter.

Let us take note, however, that the taxonomy of natural

objects has reached a relatively high degree of completion, so that it is generally possible to give a rock, a plant, or an animal a name and that this name can be referred to a reliable description that relates exclusively to it. An international system of Latin nomenclature is recognized and assures univocal reference. The published repertories (such as local, regional, or national "floras" and "faunas") also generally contain chorological information, that is, some statement on the local or even the general geographical distribution of the plant or animal. This is not nearly so standardized in format as the description itself. Therefore it is quite uneven, as you might expect, since this information does not relate to the nature of the organism but to the place or places where it is known to occur. This shift of criteria on the printed paragraph leads us out of taxonomy into ecology.

The preoccupation with geographic distribution is a very old one. It relates, on the one hand, to the origin and migration vs. the more or less simultaneous occurrence of mineral, vegetal, and animal forms, and on the other to the phenomenon of climatic tolerance.

Thus, very old Precambrian granites have been very differently altered on the cold Laurentian Shield and in the warm hills above Rio de Janeiro; the hemlocks have survived in great numbers in Eastern Asia and Western North America, are very few in Eastern North America, and have been wiped off the European continent by the Ice Age; camels, once rampant in Mexico, are now restricted to North Africa and Southern Asia.

On the other hand, the boundaries of peat-formation are known to coincide with rainfall-temperature combinations, and the same is true of the limits of tulip-tree and sugar maple; and it is water temperature that mostly checks the southward expansion of the speckled trout. The progress of invaders is likewise a subject of concern: the fungus that fairly destroyed the American chestnut and the one that is now killing the American elm were innocuous to their original hosts in Japan and in Western Europe; the small herbaceous galinsoga that lines the cracks in Montreal sidewalks hails from tropical Brazil, and the "tree that grows in Brooklyn" backyards and railroad strips so aggressively is known to the Chinese as the "tree-of-heaven". For that matter, the familiar dandelion and ox-eye daisy are not indigenous to North America either: like the majority of plants that we call weeds, they came from Europe. And so did the rat and the house-fly and the cockroach and the starling, and many pests with which we are so accustomed to live that we do not think of them as invaders from another continent. Of course the majority of us, in the Americas, are also descendants of colonists.

The geographical dimension of mineral, vegetal, animal, and human resources reveals a great deal on the dynamics of landscape. Large-scale phenomena like continental drift and glaciation put us in the presence of former land connections and the development of oceanic barriers together with climatic shifts of great magnitude, whereas regional erosion cycles explain the emergence of present land-forms and the sorting-out of coarse and fine particles that eventually stratify to generate a complex soil-system. In Eastern Canada, for instance, the violence of the Ice Age was such as to completely brush off any remnant of the Secondary and Tertiary eras, the deposits of one-hundred million years! The considerable re-shaping of the land under the impact of repeatedly advancing and retreating ice-sheets has opened new drainage ways and left the Canadian Shield and Appalachians variously scraped naked or loaded with morainic deposits, lined with old, raised beaches and terraces, and covered with soggy clay beds. The exercised eye easily follows the wake of the glacier, the tilt of the continental bedrock, the locking-in of bodies of water, the spread of bogs and marshes, the emergence of sand which is blown by the wind on the ocean shores and in the corridors of dried inland streams.

Geological maps are available for much of the surface and sub-surface of our planet, although much remains to be done.

The search for buried ores, petroleum, coal, natural gas, continues to be guided by our knowledge of the paleoclimates. Expectation is high of nickel and iron deposits on the old crystalline rocks, whereas the fossilized lush flooded forests of the late Paleozoic (or Primary) era yield coal and petroleum. There is comparatively less information on the existing landforms in the same areas: geomorphological surveys, detailing the relatively minute features of mound and vale, of cliff and promontory, of hill and scarp, do not cover a very large total surface. Such maps provide an intimate description of relief but they also promise an interpretation of the play, past and present, active or dormant, of erosion and deposit, flow and stagnation, stability and change. It is easy to see how indispensable such information is, not only to an inventory of mineral resources (including water) and interpretation of how the landscape acquired its present shape, but also to the uses that can be made of the soil.

Awareness of land-form and of soil texture is reflected in vernacular vocabulary. Baulig's repertory of geomorphology and Stamp's geographical dictionary show to what extent popular perceptions have accurately responded to real differences and to analogies, since the technical books have adopted such terms as roches moutonnées, podzol, breccia, cuesta, pahoehoe, anse, voça roca, felsenmeer, téton, to quote but a few.

Geography of the plant-cover is also rather patchy. The European countries definitely have the lead, with rather an extensive coverage, at several scales. In considering the distribution of plant-masses, it may be well to point out the basic distinction between flora and vegetation. Not a few botanists fail "to see the forest for the trees". And this is as it should be. The study of individual plants, of particular species, is a most necessary exercise. We must know the flora first, we must have as complete a list as possible of the kinds of plants that grow in a given area (a country, a county, a forest). And then, we can turn to the whole vegetation and define its composition, structure, and dynamics.

Looking first at the flora for environmental clues, we may as well consider the fauna at the same time and get confirming evidences and additional leads. Thus, in Northern Saskatchewan, some of the dominant grasses of the Prairie are widespread to the south and closely related to species of the Central European grassland, whereas others are more akin to montane meadow types; some of the shrubs are widespread in the eastern forest and others more common in the boreal forest, and others are akin to those of the scrubland in the Great Basin; the bog plants, on the other hand, fairly extend from coast to coast; as for the trees, they almost all reach down from various phases of the boreal forest; the dunes have grasses and trailing shrubs that ring the whole region of the Great Lakes; the alkaline lake-edges have a cohort of plants that usually follow the marine coastal areas; finally, many plants, deliberately or accidently introduced by man, occupy the cultivated fields and variously spill onto unmanaged land. Mammals, birds, fishes, and insects display very similar distribution patterns and geographic affinities, responding to a widespread conflict and a fluctuating resolution of grassland and forest influences, which also contain and modify the stresses of specialized habitats (such as salt lakes, dunes, and bogs): the buffalo and the deer, the larks and the warblers, the goldeye and the trout, the grasshopper and the sawfly.

A careful analysis of any given region will show how much or how little its present stock of living organisms has retained the stamp of past vicissitudes. New Zealand will bear the mark of Australian and New Caledonian migrations, of wide-low-swinging Pacific invasions, and of Antarctic connections that also extended to South America. The European Alps carry traces of ancient Himalayan outpourings, of abundant arctic extensions, and of upsurging Central European and Mediterranean migrations. South Africa reveals itself as the centre of origin of very important groups like the heathers and the geraniums (*Pelargonium*), which are mixed in with Australian and Central African elements.

Thus, an observer standing in the midst of the aspen parkland, of the Canterbury grassland, of the Swiss highlands, or on Table Mountain, will get his bearings on past invasions and emigrations, on conflicts of drought- or cold-adapted with moisture- or heat-loving organisms. He will be able to speculate on the effectiveness of wind- and animal-dispersal, on the vagaries of germination and rooting, and on the competing power of the late-comers.

Animal dispersals often closely parallel those of plants, of course. The freshwater fishes and the land birds of Eastern North America and of Western Europe often form pairs of closely related species.

Domesticated plants and animal species and their accompanying parasites and commensals (so-called weeds and pests) are living evidence of all kinds of human activities. The most obvious is pastoral and agricultural (cattle, swine, sheep, cereals, vegetables), but many are commercial or industrial, such as plant migrants along the transcontinental Canadian railways (*Collomia linearis*), Argentine thistles near the knitting mills of Southern France, or Dutch-elm-disease fungi from British freight in Quebec shipyards. These signs in the landscape will meet the eye of a naturalist, but cannot this experience be shared, and the recognition made clear to a wider audience? The book is readily opened to school-children, to "adult" classes, to the omnipresent television screen.

The latter may have helped more than any other medium to develop an implicitly analytic reading of landscape. Although many "nature" programmes are merely picturesque and often stringed upon a syrupy tale of primeval goodness, their sheer imagery impresses itself upon the mind, so that millions of people carry pictures of the arctic tundra, of the tropical rainforest, of the rosette-trees of African highlands, of the swamps of Louisiana and of the Congo, with their accompanying animal forms and cries.

The educated viewer who seeks further knowledge on landscape in the same way as he has sought to acquaint himself with individual kinds of plants and animals, does not easily encounter systematic listings, descriptions, and interpretations. Concerning vegetation, for instance, there is no world compendium and few regional monographs that compare with floras and faunas.

But vegetation of the world does not present infinite variety, if one searches for the major patterns. It could even be reduced to four: forest, savana, grassland, and desert, to which maybe scrub and tundra should be added. Even the casual traveller, reader, or viewer can find his bearings at this first level. He may even be aware that a minimum of heat and rainfall determines the world boundaries between forested and non-forested areas; that savana and grassland occur in fairly warm to tropical areas where rainfall is relatively abundant but sharply seasonal; that desert is mostly in warm and extremely dry zones but can also occur in very cold and dry ones. This over-all pattern of climatic control probably strikes the mind more than it does the total consciousness. It is almost an abstraction.

In fact, what needs to be apprehended lies well beyond such large categories, which are referred to as biochores. The question is: how many kinds of forest are there? What kind, or kinds, are to be expected in Spain, in New Zealand, in Puerto Rico, in British Columbia? Must one be able to identify all the trees according to their respective species before one can answer such a question? The answer is: no. Thus, the forests of Eastern North America are mostly of two kinds: broadleaved-deciduous and needle-leaved evergreen. Similar forests are found in Europe at equivalent latitudes. The Western Mediterranean countries, Central California, Central Chile, the Cape-of-Good-Hope, and parts of Southwest Australia have the broadleaved-evergreen hardwoods that appear to be a response to non-tropical summer drought and winter rainfall. The Congo and the Amazon basins carry very luxuriant broadleaved evergreen forests. Warm-temperate very humid areas like Madeira, most of lowland New Zealand, Southern Japan,

Southern Brazil, support less luxuriant broadleaved forest of a simplified type. In these very sketchy definitions of what plant-geographers call formation-classes, no reference need be made to maples and oaks in the deciduous forest, to spruce and fir in the needle-leaved evergreen, to laurels or camellias or Paraná pine in the temperate rainforest. The definition therefore rests upon structure, upon a bioclimatic regime that elicits similar responses in many parts of the world.

Within these greater orders of magnitude, vegetation breaks down into much finer units. Whereas the general trend of climatic adjustment may favour needle-leaved evergreen forest at latitudes that vary from about 40° to 52° in the East and above 60° in the West, Canada's boreal zone changes its composition from balsam to mountain fir, from white to Sitka spruce, and is richer in species in the West than in the East. But the various landscapes that emerge within this great forest-zone matrix have a different facies on the Appalachian Mountains, on the Laurentian Shield, in the rolling country north of the Prairies, in the mountain massifs of the Rockies, and on the Pacific Coast. The ratio of outcrop to alluvial deposits, of open water to congested drainage, of flowing stream to blocked lakes, all inhibit the blanketing of the land by forest. Regional mosaics consist of alternately fine and coarse patterns where muskeg, dune, cliff, harbour variously structured vegetation: floating mats of leatherleaf, coarse windblown grass, starry encrusted ferns, and rosettes.

It is at the level of the landscape that actual perception exercises itself, that relationship can really be witnessed, that exchanges between environment and organism can be visualized: transpiration, rooting, growth, flowering, dispersal, all fall within the ken of the observer. All senses advise him, especially if he travels on foot, of differences in plant-cover: the cool, moist air under the canopy of spruces and the dry warmth of the neighbouring goldenrod field, dark blades and sulfurous emanations in the cattail marsh. These fleeting sensorial shocks all have an ecological connotation: the resinous content of the fir, the colour and odour of the goldenrod, the liberation of gases by the muck under the cattails, correspond to defence mechanisms of the trees, to insect-attracting in the herbs, to microbiological activity in the waterlogged soil. There is much more to be known about vegetation than these warnings would indicate, for a systematic study takes in many more dimensions and I shall return to this topic in the next chapter.

I hope to have highlighted some of the interactions of knowledge and awareness, by dwelling on what is known and can be known about living and non-living features of the environment, and this will have led, in a roundabout way, to the training for education on environmental relationships. In their separate spheres, engineers, geologists, botanists, zoologists, physicians, all have direct access to partial knowledge of weather, rock, soil, plants, animals, and man. Only a very small number have ever been ecologically minded, not even foresters and epidemiologists. In order to think at all about environment, two seemingly contradictory requirements must be met: actual knowledge in some field (any field), and a capacity to relate by transfer from that field. I shall argue, later on, that there is no such thing as a good generalist who is not first a specialist. I am saying now that exposure to "information input overload" (I.I.O.), with which the multimedia confront us, has to be parried by judicious experience (again, in any field) before a valid sense of the relatedness of factors in the environment can be developed.

Much has been made above concerning the storage of environmental information and its wide diffusion. Let us not go on to plead for the rise of a generation of ecologists. Even if environmental science provides a new matrix for the unity of knowledge, this hardly makes it everybody's business, and I have no doubt that our society needs fully motivated specialists and would be rather badly off with too many generalists about. A warning has been given us in French Canada that classical studies, intended to produce well-trained minds, are a dead end in this Century, even if they were once an open road for British imperial administrators.

What sort of programme, therefore, should our society embark upon in order to make real use of knowledge related to environment and to feed the consciousness of the individual? How can sensorial perceptions of the landscape (the smog-ridden city, the flowering park, the village green, the wilderness) be deepened by increased awareness and somehow linked to the actual relationships that govern its dynamics?

As I shall concern myself, in the next chapter, with research and therefore with the increase of knowledge, I had rather dwell, as I generally have heretofore, with the citizen and the schools.

The children of today are exposed as no previous generation has been to discourses on the environment, some of which are not a little rabid. There are "ecology" games for them to play, "ecology" clothes for them to wear. For some groups, unfortunately, ecology has very nearly become a religion. Those, like myself, who have been practising ecological science for many years do not crave that much support. Our chances of maintaining the value of ecology are much diminished if it becomes political or religious.

The environmental scare, however, that is so surprisingly sudden, has very justly upset the tranquillity of educators at all levels. The disfavour under which the natural sciences have laboured, in their outdoor phases, in recent years has all but disappeared. But we cannot now go back to boy-scoutism and to National Geographic Magazine colour photography and to Walt Disney dog–boy romances. What is happening at the primary and secondary school levels is very heartening. Where the current interest in environment has been well understood by capable teachers and administrators, the erstwhile exclusive privilege of "goodness" accorded to wild nature has been rescinded and the actual environment (usually urbanized or industrialized) has become the real focus of both learning and action. This latter orientation is, of course, con-

28

sonant with the major trend of education at all levels now: learning should no longer be so passive as it once was; doing is part of learning. Skill in doing deserves as much credit as capacity to store information received.

This generation, therefore, is favoured by a better value-system, since its experience, crude as early experiences may be, is made the very matrix of systematic learning. I do not fool myself that this pattern is generalized. However, it is spreading and informing many of the curricula in our schools. Thereby it promises a better foundation for a broadening of environmental consciousness.

The need to build a new world is now a necessity, not the utopia that it may have seemed in 1914 or even in 1939. This is a work of the imagination, and imagination reaches out to hidden dimensions. I have attempted above to unveil some of the aspects of landscapes known to me and to suggest that the depth of our understanding of natural and human resources lies in such revelations. In other words, the richness of our inscapes is a preliminary to a good management of our landscapes.

Ecology was born in the realm of natural history and at a time when the divorce between the sciences and the humanities had not been pronounced. Alexander von Humboldt and Charles Darwin knew how to write, Alphonse De Candolle carried on a lively correspondence with Balzac, and Claude Bernard's *Introduction à l'étude de la Médecine Expérimentale* is a literary masterpiece as well as a scientific landmark.

The great explorers of the nineteenth century avidly bored the rock, collected and preserved plants and animals, and recorded their knowledge of place and time in words and pictures. The encyclopaedic urge for the wholeness of knowledge and for the unity of perception displayed in Humboldt's *Cosmos* stands in sharp contrast to the deliberately specialized and carefully depersonalized monographs and treatises of the first half of this century. Even strong personalities such as Theodosius Dobzhansky and C. D. Darlington resorted to stylistic devices in order to avoid the first person singular, unless and until they produced carefully labelled para-scientific papers where the man was allowed to burst through the carapace of the scientist. It is not my purpose to denounce or criticize those who conformed to this puritan lifting of science from its emotional vehicle. It is the whole bourgeois ethic that would be under trial. It is also the supposed separateness of fact and value that is being questioned at this late hour, although we have not lacked in earlier warnings from Agnes Arber, Marston Bates, and Jean Piaget, amongst others.

Watson's autobiographical chronicle of the discovery of the genetic code with all of its seemingly accidental underpinnings follows closely upon Russell's denunciation of split knowledge and Snow's analysis of the "two cultures". The time has surely come for a reconciliation of the natural sciences and the sciences of man, and indeed for recognition of authentic discovery by artistic as well as scientific and explicitly rational means. The exposure of the learning process and of the emotional leads in the pathways of scientific investigation is a necessary qualification of all breakthroughs.

Research on the environment at the present time, in my opinion, finds its best conceptual and methodological framework in ecology as formulated by field biologists and extended to incorporate the best thinking of economists. The etymological roots of economics and ecology denote more than lexical analogy, as some economists (especially Kenneth Boulding) have been pointing out for some time. It continues to depend, however, on a number of well-entrenched disciplines such as geology, geomorphology, meteorology, botany, zoology, geography, engineering, demography, economics, sociology, psychology, and history.

The recognition of a necessary synthesis involving all of these disciplines and of the urgency of harnessing teams where they are all represented must not blind us to the fact that, within each discipline, specialization must go on, must not be swamped by the requirements of generalists who will often treat its contribution as mere ancillary support. I find it necessary to build up a first panel in which I propose to consider the history and development of potentially contributing disciplines for their own sake before tackling, in a second panel, their partial or total relevance to environmental study.

In this first panel, however, I am bound to work from an ecological angle, and to point, as I go along, to the knowledge, in each field, that is most immediately pertinent to the analysis of landscape, and therefore to environmental science.

The study of the Earth and its measurement starts with a look backwards in time. The present array of mineral particles is the result of past and present interactions between the burning pyrosphere in the core of the Earth, its outer crust (or lithosphere), and the impact of the hydrosphere and atmosphere upon that crust, and indeed the blazing effect of cosmic radiation upon the revolving planet, and finally the more immediate impact of the biosphere. These forces are and have been combined in various ways. The geologist has a moving vision of volcanic upsurge, of cooling magmas, of drifting continents and rising ranges, of peneplaning erosive forces, and of vast, expanding sedimentary sheets. Natural cuts and artificial borings reveal that the same area has undergone drastic changes. For instance, Greenland and Antarctica once bore luxuriant forests, the Sahara had free-flowing rivers teeming with hippopotamus and crocodile. A full understanding of geological vicissitudes is all-important to the classification of renewable and non-renewable resources. Stable products such as iron ore, coal, and petroleum are not forming before our eyes in any appreciable (and exploitable) quantities.

We are witnesses, of course, to very similar soil-forming processes. The sciences of geomorphology and pedology are of more relevance to our preoccupation than bedrock geology. Relief and land-form distribution provide the very matrix of landscape. The physico-chemical composition of cliff, dune, and outwash is useful to know since they yield rock, sand, and clay in varying proportions to be carried or accumulated according to congested or flowing drainage patterns and alternations of erosion and sedimentation. The resulting dune or scree, the dry gravel flat, or the wet clay plain provide widely different opportunity for the effects of rain, of frost, of run-off, of accumulation of plant and animal debris. Thus peat is formed in the closed drainage of a bog and muck, in the more open flow of a marsh. The biological potential of a marsh will turn out to be very high and that of a bog quite low.

In many parts of the world, today, we have a good set of maps that show, with ever-increasing detail and on larger and larger scales, the bedrock formations and their content, the land-form investment and its monotony or variety, the emerging patchwork of soil types with their often extremely uneven distribution. Such static statements of total inventory lend themselves, in turn, to various thematic approaches related to engendering forces (the glacial cast, the late-erosive stages, the chronic eruptive blanketing), to stability and change, to narrowness or breadth of potential. Patterns of correlation emerge from the overlay of mineral content, actual or potential biomass, and management practices.

The dynamics of soil are severely constrained by a limited number of climate and relief combinations. The results of percolation, leaching, capillarity, and of physico-chemical transmutations respond to zonal regimes that we call laterization (in the Tropics), podzolization (in cool-to-cold and wet areas), calcification (in cool-dry areas), salinization (in warm and very dry areas), gleization (in cold and poorly drained lands).

The triangular interactions of climate, soil, and vegetation have been the object of a great deal of research. Some of it has been plagued by circular reasoning and uncertain assumptions. It proves very difficult to study climate in exclusively meteorological terms, soil with reference only to particle distribution, and vegetation without reference to its physical determinants. These interlocking orbits nevertheless have independent gravities and a number of botanists have successfully applied their critical powers to recognize this. A distinction has been made in the previous chapter between the various orders of magnitude. Thus large geographical areas, that are also climatic zones, are characterized by *formation-classes* such as deciduous forest in Eastern North America and in Western Europe, whereas a particular area is under the prevalence of a certain kind of deciduous forest, for instance the beech-maple forest in Southern Ontario, the maple-basswood forest in Wisconsin, the English oak forest in northwestern Spain, and the helm-oak forest in central Portugal. It is easily seen that a shift of criteria has occurred from the larger order of magnitude (the formation-class) where structure alone is implied to the smaller one (the *climax-area*) where both structure and floristic composition are invoked. Hard lines drawn on a map to show boundaries between these units are the result of much extrapolation and cannot be taken literally for they are not too frequently drawn from actual observation. With the advent of satellite photographs, however, it has been gratifying to confirm earlier generalizations from airphoto mosaics.

Accurate measurement can only take place within a landscape. It is the study of actually interacting organisms that justifies any assumption that we make at the higher levels. If we have been led to accept the idea that similar climates (e.g., in Central California, in Spain, and in Central Chile) have elicited similar responses in the structure of upland vegetation, this hypothesis concerning formation-class has to be supported by actual knowledge of convergent evolution, of morpho-physiological adaptation, and of mechanisms of migration. In the next lower frame, that of the climax-area (beech-maple forest in Southern Ontario, oak-hickory forest in Eastern Illinois), it is our accumulated knowledge on vegetational change and on the tendency to convergence that upholds the usefulness of a particular plant-community as an indicator of maximum stability within the present climatic cycle.

The laws that govern the mutual adjustment of plant species sharing a common habitat are complex and highly subject to regional staggering. I can only hope, in the present context, to illustrate some of the processes that govern the ecology—I should say the economy—of these interactions. I am convinced that it is very important to do so, if only to emphasize the frequent shift from competition to co-operation which seems essential to steady renewal of resources.

For a long time, plant-ecologists greatly stressed the power of competition among plants and gave short shrift to co-operation. A historian of science may well question whether such an emphasis was unconsciously supported by puritan-capitalistic culture, and look at this development in the same light as the debate on heredity of acquired characters where ideological inroads have often been all too evident. However that may be, a more comprehensive approach consists in investigating the devices wherewith different species share the resources of the environment. *Sharing processes* operate simultaneously or successively. I shall offer two examples.

The maple forest, in early spring, allows full sunlight to reach the soil, and a large number of herbs display an overlapping phenology, with rapidly alternating budding, leafing, flowering, and even fruiting; moreover, their underground parts are buried at depths that are remarkably constant for each species. The provision of light, heat, and moisture from air and soil, the organic and mineral nutrients segregated in several soil layers therefore offer a finely graduated scale that favours differential intake. The oncoming summer raises and stabilizes the heat level, but the expanding leaf-canopy reduces light and direct radiation and maintains somewhat constant moisture. Many of the spring herbs then disappear from the surface and their underground parts go dormant. The summer herbs rise in the shade, rather scattered and much taller in comparison to the earlier occupants, and many of them are conditioned to late-flowering by the shortening of daylight hours. A close look at this biological clockwork reveals a well integrated pattern of tapping and transforming and return by decay. In fact, its harmony assures long-standing renewal and stability.

In a Western American Prairie, a similar rotation of exploitants can be seen, with the grasses growing up from soil level, among the protective mulch of the preceding year's dried stems and leaves, each species successively thrusting its flowering spikes that turn to gold in the summer, and later to brown. Meanwhile the forbs (or broadleaved herbs), having formed rosettes on the ground in the growing shade of the grasses, sprout leafy stems that undergo the sexual crisis which releases their flowers under the impact of longer nights in August. Underground, the finely cut fibres of grass roots reach to great depths, whereas the coarser, often tuberous parts of the forbs lie nearer to the surface. Considered at close range, the Prairie is just as seasonal as the forest and possibly as complex also. The dramatic changes of spacing, of texture, and of colouring are, in both instances, signals of a shifting resource pattern for food- and shelter-seeking insects, birds, mammals, whose seasonal behaviour responds to thresholds of availability.

There are not nearly enough local monographs recording in fine detail the march of the seasons and detecting the beginning, the maximum, and the end of each process in each commensal species. There are enough, however, to document the relative uniformity/diversity, efficiency/waste, and stability/change in environments as sharply contrasted as the forest and the grassland, the Arctic and the Tropics, the marsh and the desert.

A preoccupation with change has always run very high in the minds of students of vegetation, whether they sought records in bogs and in lakes of the effects of warming climates on retreating ice and advancing waves of plants, or set their sights to the minute and less speculative observation of drying lakes, shifting shorelines, or moving sands. Bogs and dunes, so abundant in Northern Europe and North America, continue to fascinate naturalists and to lend themselves to textbook demonstrations. It was chiefly on these environments that the Chicago School, at the turn of the Century, developed the principles of plant-succession, although earlier observers (and notably Kerner von Marilaun in his study of the Danube Basin) had foreshadowed them. We must never be surprised that a good idea gets overworked, and it is evident that this one did to the extent of earning discredit among the researchers who became more intent on detecting arrests than continuity in the flow of change.

I do not propose to review this controversy, now over-shadowed by more newly emerged issues. I shall only retain some of the principles involved, for I am convinced that their application to areas as apparently far-removed as urban development is of great significance.

The fact is that all vegetation is not stable and that a predictable replacement of one kind by another is the result of two basic phenomena. The first is that occupancy of a site by most plant-communities more or less gradually alters the resource base; and the second is that many plants are physiologically unable to maintain themselves under the conditions that were favourable to their establishment. Evidence for both points is readily seen where unoccupied mineral ground is invaded by weeds that will soon have incorporated organic matter where there was none. A new threshold of fertility is then reached, opening the way to species with a higher requirement. If the newcomers are also better equipped to tap air, water, and nutrients, they are bound to displace the pioneers. The latter therefore prove themselves unable to exploit their own investment. The ecologist's task consists very largely in defining what the resources are at points A, B, C, D in a series of such replacements. Thus, in a succession that ranges from bare, overdrained sand to forest through a grassy sward, a lush prairie, and a shrubby savana, the *critical factor* may first be soil-water, then organic content, and later decreasing light at soil level, but it is eventually the very capacity of the plants themselves to fulflll their cycle.

The change is gradual, but the rate of change frequently undergoes either acceleration or lag, even to the point of stagnation. Each line of succession is typified by the physical nature ′of the site and its yield of resources at its initial point, which may be limestone rock, dune sand, saltmarsh, etc. The constraints of pioneer conditions are the most stringent upon the earliest invaders and it is the buffering effects of further consolidation that permit the beginnings of convergence. Thus a goldenrod prairie may well develop on what was originally either a limestone gravel or a sandy flat.

In a large number of areas, from the Tropics to the Arctic, it has been possible to draw a network of such successions from many actual initial points and to ordain the entire regional vegetation in a coherent scheme of convergence towards a hypothetical climax. Across such lines of replacement are figured arresting blocks that impede further change, such as periodic floods or fires or very slow geomorphic process.

Animal ecologists have often paid even more attention to vegetation dynamics than have plant ecologists, for they cannot do without a plant-mass matrix upon which to project animals, themselves defined as to kind, number, and movement. Whereas plant ecologists, especially plant sociologists, have only too often sampled a particular plant-community (white-pine forest, cattail marsh) in widely disconnected stands, such a course is hardly open to the animal ecologist whose quarry is almost always moving across entire landscapes.

The preoccupation of the animal ecologist, even if he is experimenting on the intricate metabolism of a given species or population, is never far from the detection of spaces where the requirements of one or more animals can be fulfilled. It will turn out, for instance, that some birds will feed in one habitat, will breed in another, and will rest in a third, not to mention their seasonal migration out of the area altogether. It remains to investigate whether the field is chosen primarily because of light or wind conditions or because of the presence of specific foodstuffs, whether the forest offers more constant temperature or greater safety, and so on. The investigator of animal behaviour under natural or semi-natural conditions depends upon a repertory—often designed by plant ecologists—

that defines the resources base, at least implicitly, in all its useful dimensions.

Here again, as with associated plant populations, the phenomenon of *sharing* appears. Of the twenty or more species of birds that occupy a shrubby field, which ones are nesting, feeding, resting? Which ones use only certain plants? Which ones are insectivorous and omnivorous? Which ones are residents and which ones are migrants?

In most areas the answers to these and to many parallel questions are well known only for some groups. Birds, for instance, since bird-watching is an antique sport as well as a full-fledged science. There is probably a better international accord on bird names and on retrieval of bird information than there is on any other major group, although Vertebrates in general have received a large share of attention. Invertebrates, on the contrary, are not nearly so fully inventoried as to species, and there are not too many areas for which comprehensive quantitative data are available. Stomach contents of fishes, birds, and mammals have given us some very useful clues, however, so that for a number of habitats we have an accurate picture of well-calibrated food-chains.

Geology, botany, zoology form the backbone of the sciences of nature, although chemistry and physics have to be added to the natural sciences. Their acquisitions through the labours of research workers have been summarily pinpointed above in special relation to environmental science. It now remains to shift to the panel of the sciences of man (some with a considerable natural-science content) and to focus them upon the landscape which has been envisioned so far, in this chapter, without its man-made changes. It is, of course, the very purpose of agronomy, engineering, medicine, sociology, and economics to reveal just how man manages himself and his environment.

The stamp of man upon landscape will be considered in fuller detail in the next chapter. For the moment, however, let us see the implications of the relays of energy from the geological crust to the plant-cover on to animal transformations, and ask ourselves how human investment has meted out a variety of management practices.

Investments are made in the landscape wherever and whenever resources are stored that have a potential deferred use. Buried minerals lay untapped as long as a physical process such as erosion does not expose them or so long as a biological process such as root penetration does not reach them. It would, however, be an abusive extension of the term to call all latent resources "investment". It is best to apply this term, if we use it to denote an ecological phenomenon, to storage of materials or construction of artifacts that have a long-term usefulness for the maintenance of a certain condition or for the cycling of one or more resources. Thus, a biennial plant like the evening-primrose or the mullein produces the first year a vigorous rosette of leaves whose function it is to store in the heavy taproot a large quantity of reserves that will be used up at the flowering and fruiting stages the second year. Many bulbous perennials have a similar habit of deferred utilization. Hibernating groundhogs and bears store fatty tissue to maintain their reduced metabolism and migrating birds do likewise to fuel their long flights.

This kind of investment affects the individual alone. Another kind may be seen where other participants in the community are destined to use the stored resource: for instance, large quantities of sap, sugar, starch, and cellulose accumulated by woody plants are used by parasites and epiphytes and eventually, when the tree dies, they are digested by bacteria and ploughed back into the soil.

In the animal world, both of these processes are quite common. Squirrels and many other mammals are given to hoarding. Almost inevitably the caches also serve the purpose of proximal dispersal and many a "forgetful" squirrel has actually sown hickories and walnuts. The same squirrels also gather leafy twigs with which they construct nests for their young high up in the trees. Birds, of course, are the great architects of nest-building, from the neat little baskets of the warblers to the hanging hammocks of the orioles. Bank swallows and petrels, however, are burrowers whose recesses are not unlike those of foxes or woodchucks. Raccoons, on the other hand, delve into dead or decaying tree trunks. No doubt the fine art of tunnelling reaches its perfection in gopher "cities" and in termite mounds.

It is always of the greatest interest, ecologically, to identify the materials bored into, readjusted, imported, or secreted by animals in their construction of shelter. The manufacture of wax by bees, of paper by wasps, the plastering of mud by swallows, the lining of their dens with herbs by foxes all affect the energy transfers in a measurable fashion. But further steps are taken in the division of labour among castes, and the cultivation of fungi and the domestication of aphids by termites is a veritable prefiguration of agriculture and herding. Possibly an even greater impact is made by beavers whose carefully engineered dams set the water at a constant level, thereby determining what kinds and quantities of plants and animals will prevail.

Such a re-orientation of mineral, vegetal, and animal forces in a sizeable portion of the environment by a wild animal is perceived as "natural", whereas the cutting down of a forest and ploughing of the soil by man is said to be "artificial". If it be argued that the difference lies in the use of tools, it must be said that many animals also do: the sea-otter uses rocks to crush the shells that contain its food; one of Darwin's finches digs grubs out of the bark of trees with a cactus spine; monkeys are also known to manipulate sticks and stones.

Unquestionably, however, human investments superimpose themselves much more powerfully upon the mineral assortment, the plant-cover, and animal exploitation. I shall give this further attention in the next chapter. However, in my attempt to draw the picture of our growing knowledge of environment from its origin in the natural sciences to its psycho-social implications, I find it imperative to consider the "role of man in changing the face of the earth". This was the title of an epoch-making symposium in 1952, in which it was proposed to take stock of the many ways in which natural landscapes have been affected by man. It somewhat belatedly inventoried many kinds of transformation, harnessing, and substitution, very often under the labelling of "damage", "interference", "spoilage". Two more symposia (1965 and 1971) were to readjust this perspective to a rapidly shifting vision of man's place in the natural world. I was a contributor to the 1965 conference and, at that time, developing some of the themes that I am concerned with here and now. Without attempting a full review of the contributions of geography, engineering, economics, sociology, psychology, I shall point out some parallels to the acquisitions of the earth sciences and the biological sciences, in the hope of detecting points of convergence and of identifying facts and interpretations that are useful to an emerging human ecology and a broadly based environmental science.

In French universities, departments of geography are still tied to Faculties of Letters, maintaining their traditional links with history and political economy. This has not prevented the French school of human geography from developing a methodology and from aiming at objectives that were being missed on the one hand by geologists who brushed aside the human factor and on the other hand by sociologists whose concerns were too philosophical. It may well be that Jean Brunhes and Vidal de la Blache were more direct precursors of human ecology than their Anglo-Saxon counterparts, steeped in a more analytic sociology. Unless, indeed, it is the anthropologists and ethnographers who first staked the claim.

Instead of searching out the historical priorities, I do believe it is more pertinent to my present task to identify the sources of information and the conceptual schemes which are most pertinent to my topic, as I have tried to do above.

An ecologist, culling the field of human sciences for data on environment, gets some of his leads from geographers. It is they, and not the ecologists, who have striven for a synthetic view of landscape. The classification of weather types, the nomenclature of land-forms, and the elaboration of functional maps are the result of systematic spatial investigations not undertaken by anyone else. Valuable as historical and economic studies are, they have traditionally been etched against a very softly focused background, when they have not been virtually devoid of recognizable spatial dimensions altogether. Many demographic and sociological studies also tend to a mathematical abstraction that remains to be fitted into a material environment. I do not wish to be too critical of historians and social scientists who do not or cannot cast their findings into a clearly visible environmental mould, for their data and interpretations remain very valuable for the ecologist, as much so as the specialized data provided by the geologist or the geneticist whose concern does not extend to the entire landscape.

This type of work, again, is best done by the geographers. Thus, a land-use survey of the kind that we have going in Canada at this time gives a running inventory of spaces devoted to the varieties of human exploitation: wild land, agricultural, residential, industrial, urban, and the many subcategories that emerge regionally.

Fields of applied science, which have been largely under the control of professional corporations at the level of study as well as of practice, have a great repository of fact and experience that the environmentalist can use. Agronomy, fisheries, forestry, medicine, and engineering have developed ever more efficient techniques for dealing with their objects, thereby meeting the challenges of environment, and modelling the landscape. Exploitation of mines, forests, streams, and soil; building of roads, shelters, factories, cities; growth of technology and protection of health are impossible without a considerable number of experiments and inventions, of trials and successes.

It is therefore to these professional fields that the ecologist

43

must also turn if he hopes to define the habitat of man in as satisfactory a fashion as he has done for the moose, the beaver, or the honey bee. Changes in types of architecture, shifts in building materials, new means of communication, to him are not primarily of historical interest; the ecologist spells them out in terms of energy flow and assesses them as indigenous or imported and as efficient or inefficient.

This brings us back full circle to our preoccupation with the psychological dimensions of the landscape, with man's projects vs. his accomplishments. The growing number of choices that his increasing power permits is illustrated by the unfolding of the sciences that have supported industrial growth and urbanization, including the more recently emerged urban and rural planning. But the traditional technologies also lend themselves to a minutious scanning. The origins and the whole history of agriculture, fisheries, medicine, and engineering are fraught with interacting scientific discovery and mythical impact. The relation of man to maize and rice, to shellfish and whale, to health and sickness, to shelter and communication, is impossible to map out if we cannot gain access to the objective facts of resource availability on the one hand and to the mechanisms of psychological urges and constraints on the other.

If Indians will not consume the meat of cattle, if French Canadians refuse to eat mutton, if Hebrews will not use fuel on the Sabbath, if wealthy people will not use public transport, how do these psychological halts figure in an ecological scheme? Undoubtedly as levers that check the normal flow of energy. Blockages of this sort have effects in resource cycling that are quite similar in their strength to flooding, to food shortage, to excess of combustibles, or to congestion of transport.

Having gone this far, the ecologist will have recognized man as part of nature, will have undertaken an inventory of what leads to the production of resources, will have classified landscapes in terms of their natural and superimposed man-made patterns. In order not to break the continuity from the natural to the human sciences that contribute to a rational estimate of landscape, an ecological framework must be designed.

The notion of ecosystem, long implicit in the thinking and methodology of ecologists, has been more clearly formulated in the past forty years or so and has finally given rise to an increasingly precise and quantitative methodology. Drawing together the accumulated facts and theories of the sciences bearing upon environment will be the task of a whole generation. This new Renaissance, as Max Nicholson has called it, can only take place, like the previous revolutions in man's history, if man holds a different view of himself than he has had until now. There is much evidence to that effect in our day. The opportunity seems to avail therefore and the means are also at our disposal

I propose to pick up this thread while considering the modalities of man's escalating power, and his efforts to design and plan under a rapidly shifting motivation.

The ecosystem is a more or less closed environment where the resources of the site are cycled by a biomass of plant and animal populations associated in mutually compatible processes. A careful analysis of this proposition is a requisite to the applications which I intend to make. Many earlier definitions have led to this one. Tansley in 1935, Lindeman in 1942, Odum in 1953, Duvigneaud in 1963, had provided the now classic representation of a layered triangle or pyramid within which the functions of resource cycling by green plants, herbivorous and carnivorous animals were charted. The shape of this pattern was intended to reflect the diminishing numbers of species (and individuals) involved in the successive episodes of cycling. Whereas this design remains valid, I have recast it somewhat differently in order to set up the mineral resources at the base and to account for the controlling factors at the top, including the psychological. I shall try, therefore, to gather many of the threads that have been woven in the previous chapters and to draw them together into a coherent image of the ecosystem.

But first, an over-all placement of the activating forces and a definition of terms. In any given ecosystem, the *resources* are cycled by *agents* and this is accomplished through *processes* that are typical of each *level* in the transfer of energy, or *trophic regime*. The result of the agent's action upon the resource is a *product*. And again products are characteristic of the trophic level at which they emerge.

This latter consideration allows the recognition of six such levels: mineral, vegetal, herbivorous animal, carnivorous animal, investment, and control. The diagram on next page is one way of showing the cycling involved. In the central stream of energy flow, the minerals are taken up by green plants, these in turn are consumed by animals that serve as food to other animals, and the next level indicates investment and storage, whereas the top level is the zone of control. This notion is derived principally from the observation of power on the part of an agent to operate levers at virtually all levels of the ecosystem. It is also related to an implicit foresight in the continued management. Vernadsky (1945) and Teilhard de Chardin (1955) have coined the expression noösphere to encompass this intrusion of will (or of mind) in nature.

Relays of energy, capture of resources by agents, however, do not always operate in this linear order. That is why, on the left of the diagram, an upward movement allows the feeding in to each higher level from all lower levels; that is why, on the right, feedback from each higher level to each lower level is allowed. Thus far, the ecosystem as a whole would have no inlet and no outlet—a very rare situation. More often, as indicated on the diagram, some influence at some level would reach into the ecosystem (arrows on the left) and some products would emerge from it (arrows on the right).

To illustrate these definitions in a concrete way, let us consider such an ecosystem as the Laurentian sugar-bush. A vertical relationship can be envisaged whereby the relays occur from I to VI in the following way.

At level I, *mineral* resources are provided by heat, light, and energy radiated by the sun, by rainwater and groundwater, by weathered bedrock and washed-in materials; the interaction, physical and chemical, past and present, of non-living forces has resulted in a layering of particles at various depths in a soil that displays considerable stability and renew-

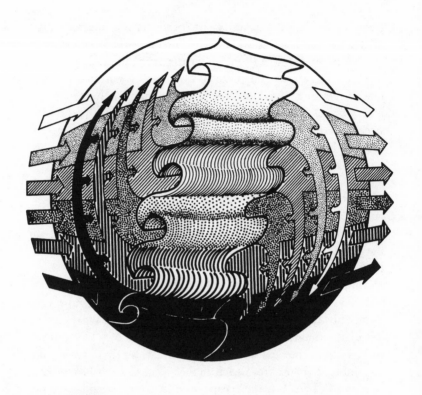

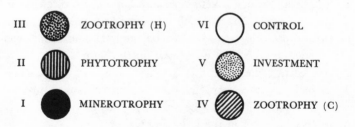

III	ZOOTROPHY (H)	VI	CONTROL
II	PHYTOTROPHY	V	INVESTMENT
I	MINEROTROPHY	IV	ZOOTROPHY (C)

A scheme showing a projection on the six trophic levels of the mainstream flow of energy (central part), the supply of resources (left part), and the reinvestments (right part), as well as the import (left margin) and export (right margin).

ability under the periodic impact of centuries of seasons. Insolation, evaporation, sedimentation, pedogenesis (soil-formation) are the principal processes.

At level II, the *vegetation* (trees, saplings, shrubs, herbs, mosses, fungi, and microbes) directly taps the energy of the sun (photosynthesis), uses light, heat, water, and mineral as well as organic elements in air and soil (assimilation), thence transforming them into sugar, starch, cellulose, and various much more complex substances which circulate and settle within the organs in the course of differentiation.

At level III, *animals* of many kinds (insects and other invertebrates, amphibians, reptiles, birds, and mammals) feed upon roots, bark, stems, wood, leaves, flowers, fruit, and seed (herbivory or phytophagy), turning plant tissue into scale, wing, claw, fur, blood, flesh, etc. This morphogenesis (emergence of specific form and function) is even more complex and less reversible than that of plants.

At level IV, other *animals*, many of which consume no vegetable substances, prey upon the herbivorous kinds, and they, in turn, also produce complex animal substances. At this level, the relays may become very numerous: thus an insect may be eaten by another insect, which in its turn will be consumed by a woodland vole and the latter by an owl. Such *food-chains* may be long or short, but are quite characteristic of the ecosystem within which they develop. Animal activity at levels III and IV includes many operations that influence the forest ecosystem: migration, burrowing, dispersal of seed, etc.

At level V, a number of *investments* appear. They consist of resources that do not enter into the current cycle or else appear to be geared to long-term functions. Such are the increment of annual woody layers in the trees, accumulated reserves in the roots and underground parts of many plants, stored nuts and tubers in the dens of burrowing mammals, fatty tissue in the hibernating woodchucks. Excavated tree trunks harbour several generations of woodpeckers and rac-

49

coons; birds design and defend "territories"; man has also drawn pathways and erected a cabin equipped with stove, boiler, and evaporator.

At level VI, one searches for over-all *controls*, which are not especially apparent in this instance, except for some selective cutting by man. The truly dominant influence is climatic, in the sense that the stability of the maple forest is geared to the equilibrium of soil and climate.

Such a relatively simple ecosystem, with a light harvest by man of easily renewable plant products (wood and maple sap), is characterized by a stable mineral level, and exceedingly active and well-balanced plant and animal levels. It is also remarkably self-contained, since virtually no minerals or vegetables enter it from the outside and very litle plant-material is exported, and, save for migratory nesting birds and nomadic predators, the bulk of its animal life is sedentary.

At the other end of the ecological scale, we can turn to an urban ecosystem, focusing our attention, for instance, on downtown Montreal. It is immediately evident that mineral, vegetable, and animal materials are all derived from other ecosystems: roads and buildings, tools, machinery, paper, furniture have all been imported, and save for rare patches of lawn, a few trees, potted plants, all vegetables come from outside, as do fish and meat marketed and consumed by man. Levels I to IV are thus virtually inactive within the ecosystem itself. On the contrary, the investments (level V) loom very large, and possibly larger are the controls (level VI). These controls, expressed in commercial, administrative, economic, political, religious, and cultural terms, not only determine the quality, quantity, and timing for the import of stone and steel, vegetables and meat, for the building and renting of housing and office space, civic, commercial, cultural, and religious services, but they also extend their impact to mining (level I), agriculture (level II), hunting (level III), slaughtering (level IV) where the products of other numerous and far-lying ecosystems are potentially very productive.

I will not attempt, for the moment, to draw the full gamut of intermediates between such extremely different ecosystems, but I will take my bearings several times again upon them, as I attempt a review of man's escalating influence on landscape, for it does follow a pathway from autarky (or self-sufficiency) to complete dependence.

But let us recapture our definitions before escalation is considered.

Resources are those elements that variously serve the cycling processes; some of them are mineral (light, air, water, soil), some biological (plants, animals), and some are services (cleaning, information).

Agents are living organisms that have the ability of engaging in the various processes by intake, transformation, storage, channelling, and conveying of resources.

Processes are the mechanisms whereby the resources undergo all and any kind of change or transmutation; they all imply energy flow. Such are weathering, pedogenesis, photosynthesis, drinking, predation, damming, transmission of electricity, marketing, stock-exchange speculation.

Products are the result of processing and are either stored or lost or invested as resources for potential recycling.

Trophic levels are more or less clearly stratified as the cycling processes carry the resources from one state to another, e.g., from the mineral to the vegetable to the animal. Thus the levels are minerotrophic (weathering), phytotrophic (photosynthesis), zootrophic (phytophagy and predation), noötrophic (damming, urbanization).

Thus the *structure* of any ecosystem (a pond, a forest, an orchard, a suburb) is characterized by its *resources* and the relative dynamics of its predominant *trophic regimes*. For instance, the desert and the arctic tundra described in the first chapter are minerotrophic, the rainforest and the prairie are phytotrophic, and the seal-islands and the African savana are zootrophic. A city, on the other hand, is noötrophic. It is important to point out that the products of one trophic level (for instance vegetable tissue) become a resource for another trophic level (herbivorous animals), and indeed for other ecosystems.

The effectiveness of the processes (photosynthesis, respiration, digestion) determines the quality and amount of production, therefore the *productivity*. In that sense, the vegetable product is called primary production; the animal matter resulting from consumption of plant material is alluded to as secondary production. Concurrently, the eating of plants by animals is referred to as primary consumption and of animals by animals as secondary consumption.

The interlocking cycles are best considered by a detailed analysis of each ecosystem. Many such studies have been made. The International Biological Programme has allowed us to reach a higher level of expertise and to develop greater confidence in this respect. The uniformity of procedure and of basic conceptual agreement, which is nearly impossible on the political plane, is readily achieved in a well-defined scientific co-operation. The methodology proposed in a series of handbooks has been adhered to by the biologists of the world, and their results are therefore comparable. We now know a good deal concerning the primary and secondary productivity of many ecosystems. We can compare the deciduous forests of North America with those of Western Europe, the Soviet Union, and Japan; we can evaluate the relative efficiency of freshwater bodies in tropical, temperate, and arctic zones.

The IBP has concerned itself primarily with so-called "natural" environments. There are two reasons for this. First, is our need to understand how the species of plants and animals have adjusted to their environment and to each other within the sway of natural selection and geological change. Previous knowledge of this kind is a necessary background, as I shall argue, to any estimation of man's impact. The second reason is historical, and therefore partly accidental: the generations of ecologists who were trained before the initiation of the IBP did not readily take man's "interference" in their scope. Their eye and their mind were usually focused on nature minus man.

I belong to that generation myself, and although I may claim to have emerged earlier than some from this constraint, I have no mind to put the ecologists of the Fifties and Sixties on trial, inasmuch as I adhere very strongly to the first reason mentioned above. We can only stand in awe of the amazing amount of work that was done by the IBP ecologists (and often with very restricted means), and be grateful that we have made such a great step forward, and, if I may say, upwards.

The wind-up of IBP will take several years, and the final summaries and the summary of summaries will serve as landmarks for years to come.

Meanwhile, in international circles, and especially at UNESCO, the resonance of the environmental crisis has resulted in an increasingly ecological definition of resource management. The first conference on "Man and the Biosphere", now well known as MAB, was held in Paris in 1967. Several further conferences have brought to a sharper focus the objectives of this new international venture, which will profit by the precedents that the International Geophysical Year (IGY), the International Biological Programme (IBP), and the International Hydrological Decade (IHD) have set. It was my privilege to participate in two of these sessions, in Paris in the fall of 1969 with the natural scientists, and in Helsinki in the spring of 1970 with the social scientists.

On this latter occasion, I presented my views on "Ecology and the Escalation of Human Impact", and I would like to re-engage in this perspective now.

Man's growing exploitation of environmental resources has followed the devious path and erratic tempo of other evolutionary and behavioural developments. In other words, different human populations (ethnic groups) have followed dif-

53

ferent itineraries and it will be found that most of them, in reaching a new (higher?) level, have allowed the persistence of earlier ways of life. It can well be argued that a diversity of exploitative activities is the more harmonious, if not the most productive.

At all events, it seems useful to isolate the "steps" as separate accomplishments and to seek out the ecological and cultural processes that are at work in each one of them. Ultimately, regional landscapes should be searched and analyzed to detect the relative importance and weight of the various operations.

Five major phases can be recognized and each one is initiated by a revolution. This term needs to be sharply defined in an ecological context. It refers to a major discontinuity in the rate of change, which is triggered off by a new access to resources hitherto untapped and therefore to a new productivity potential.

Should we consider one by one all six levels described above, authentic turnovers can be detected within the space-time operation of each regime. Thus, at the mineral level the geology textbooks have traditionally recognized five eras, each of which is marked by a major upheaval caused by tectonic movements (mountain building, continental drift) and/or by widespread climatic change (glaciation). Thus late-Algonkian, late-Permian, late-Cretaceous, and late-Pliocene all bear the traces of spreading ice and they usher in a new era. It lies very close to my present purpose to draw attention to the following fact: as the geological strata variously overlie and compenetrate each other, they have re-used the same materials and reassorted them. The end result is an increasing differentiation of the mineral matrix and therefore an ever-growing variety of shapes, forms, and states in which assimilable mineral resources are to be found. Likewise, the geologists' "revolutionary periods" (maximum differentiation of relief and climate) are richer in air–soil contacts (and therefore in the variety of resulting soil-types) than the "normal", more equable periods.

At the second, third, and fourth (biological) levels, plants and animals manifest parallel but not necessarily simultaneous discontinuities. If we look at the calendar of first appearance of major groups of plants and animals and consider them in their taxonomic status, the most significant steps would seem to be: unicellular life, multicellular life, molluscs, fishes, vascular land plants, amphibians, insects, birds, flowering plants, mammals, and anthropoids.

I cannot dwell on the ponderation of each of these accomplishments in the diverging evolution of living organisms, except to single out the increasing number of ways in which they become equipped to tap ambient resources and to cope with diversifying environments. No doubt, as Teilhard de Chardin would have it, the steps of corpusculization and phyletization are of greater magnitude than the acquisition of flowers by plants and of backbones by animals. From an ecological point of view, however, these later developments are of more interest, and the great contests of the ferns, the gymnosperms, and the flowering plants for the dominion of the forests, of the reptiles and the mammals for dominance on land are more readily perceived in terms of their capacity to monopolize air, soil, and water or vegetal and animal food.

We are therefore led to visualize the contact of the lithosphere and atmosphere with the biosphere in terms of revolutions such as vascular development in plants and animals, air-breathing and land-habit again in both kingdoms, rooting in plants, and locomotion on land and in the air and water in animals. Each one of the major transformations gives access to new resources, concomitantly with an advancing differentiation of structural and visceral devices that allow a greater efficiency of the metabolism.

At level V, *investments*, as previously defined, cross parallel thresholds. For instance, the flowering plants have developed capacities unknown to ferns and other earlier groups: they can accumulate layer upon layer of wood, year after year; they can perpetuate their kind in cold or dry climates by highly resistant dormant seed. This kind of avoidance of unfavourable environmental conditions probably reaches its highest expression in migratory birds that invest in body fuel to be used in transcontinental flight. Building patterns develop first among the invertebrates (molluscs, insects, spiders) and reach a high degree of perfection "at the hands" of birds and mammals. The main thresholds, therefore, are: body storage and insulating substances in plants and animals; external food storage in animals (caches); architecture of abode (tunnels, galleries, nests, huts) by animals; utilization of implements and construction of engineering works (dams) by animals.

This latter feature, in truth, belongs to the next higher level (number VI), that of *control*, or the exercise of power upon the habitat as a whole. Termites, oysters, guano birds, beavers, transform large segments of the environment, re-assorting mineral particles, redistributing aeration, eliminating and introducing an old and a new flora and fauna, orienting the ecosystematic forces towards certain products mostly of benefit to their own kind.

It is by taking my cues from this summary run-down of innovative occurrences at each of the six levels in the ecosystem that I detect five phases in man's escalating power over his environment. My description of these processes should be substantiated by better anthropological and sociological knowledge than I can presently command, and I shall be only too happy to obtain the attention of specialists in these fields. But I have already ventured to draw an ecological infrastructure to the "phenomenon of man" and to plot his role in "changing the face of the earth" in environmental terms (1969, 1970, 1971), and will use this scheme once more, with minor modifications. This is summarized in the table on next page. It is based upon eight distinct steps or *stages*, which are grouped into five *phases* (A to E), and transitions are initiated by *revolutions* (I to VI). Phase A is primeval,

PHASE	STAGE	IMPACT	REVOLUTION
E. CLIMATIC-COSMIC	8. EXOBIOLOGICAL OUTBURST	escape from gravity	
			VI. COSMIC
	7. CLIMATIC CONTROL	atmospheric alteration	
			V. CYBERNETIC
D. INDUSTRIAL	6. URBANIZATION	agglomeration of construction	
	5. INDUSTRY	substitution (mineral), fabrication	
			IV. INDUSTRIAL
C. SETTLEMENT	4. AGRICULTURE	cultivation, selection, substitution (biological)	
			III. AGRICULTURAL
B. NOMADIC-PASTORAL	3. HERDING	grazing, fire, erosion, transhumance, propagation	
			II. DOMESTIC
A. PRIMEVAL	2. HUNTING & FISHING	ablation, displacement, fire	
			I. IMPLEMENT
	1. GATHERING	ablation, submission	

and contains Stages 1. gathering, and 2. hunting; Phase B is nomadic-pastoral, and contains Stage 3. herding; Phase C is settlement, and contains Stage 4. agriculture; Phase D is industrial, and contains Stages 5. industry, and 6. urbanization; Phase E is climatic-cosmic, and contains Stages 7. climatic control, and 8. exobiological outburst.

I can hardly state too strongly my own reservations about any such scheme, which is useful in giving us our bearings, but can hardly be applied literally. Let it be emphasized that it is based upon *processes* capable of directing resource-utilization by man. It should be obvious that many landscapes reveal the mark of the concomitant influence of many of these processes, such as a mosaic of wild, agricultural, and industrial land-use. It cannot even be implied that the chronological order of appearance of these interventions has followed the order of 1 to 8. It is evident that some industries have been set in completely virgin land, without benefit of a previous pastoral stage. Likewise, agriculture, as Carl Sauer (1952) has insisted, has not necessarily emerged from herding. These cautions notwithstanding, I do believe that the eight stages grouped into five phases offer an actual reflection of man's growing power and allow a placement of forces in the landscape which is an indispensable requisite to further analysis.

As we consider the altering man–landscape relationship, we shall ask ourselves, at each stage in this progression, what is the *taxonomy*, what is the knowledge and classification of objects (mostly plants and animals) upon which this relationship functions? We shall also want to know of what kind is the *ecology*, or knowledge of environment. Finally, we can appreciate the *impact* of man's management.

Phase A (*primeval*) is characterized by *submission*: where man creates little or no "disturbance" in the landscape, where the application of his powers scarcely exceeds that of other animals. Virgin land, so-called, is a point of departure where the free play of ecosystems as defined above encounters no resistance and no marshalling by man.

Thus the first step is taken by the *gathering* society (**Stage 1**) which inflicts little disturbance upon the environment. The killing-off of a few trees, shrubs, and herbs does not sensibly alter the structure of the vegetation; the ablation of foliage may tend to diminish the vigour of certain species; the harvesting of large quantities of fruit, the stealing of turtle and bird eggs, of honey, oysters, insect larvae, and fungi may also tend to restrict the abundance of a species, although it also serves to disseminate it!

At this point, a first revolution, tool-making or implement development occurs, leading to **Stage 2**, inasmuch as *hunting* and *fishing* require a minimum of technology in the form of tools or implements and devices. The requirements of *hunting* are of a higher order than those of gathering. The taxonomy of the hunter's targets may be a simpler one, in that it involves fewer objects, but it has fewer morphological and more physiological features (e.g., quality of flesh and viscera, speed and wariness, bulk and danger to hunter). The ecology is closely geared to favourable and unfavourable phases of life cycle (mating, breeding, rearing, migration of ungulates and birds); to gregariousness (and whether specific or seasonal); to abundance (and whether specific, seasonal, or cyclic); to habitat (and whether narrow or wide) as determined by topography, cover, feeding and drinking habits. The resulting strategy of locating and stalking game is therefore based on an eventually broad repertory of natural history and calls for appropriate devices (axes, spears, snares, nets, etc.) which were invented or imported (the complex harpooning of the Eskimo, the poisoned arrows of South American Indians, the boomerang of Australian aborigines, etc.). The use of fire deserves special mention since it has shaped the very structure of vegetation in areas long used by hunters and since it has also served as a selective force in the evolution of woody plants (thick bark, fireproof diaspores) and of grasses (regenerating rhizomatous versus aerial-stemmed species). Selection within the hunted and the hunting populations themselves is not negligible: the well-known effect of constant, if not heavy, predation upon animal species is to conserve health and fleetness; and upon the human population some advantage accrues from fleetness also, and the capacity to plan and to lead. In this man likens himself to the lion or to the wolf.

At the hunting level, therefore, there is a certain obedience on the part of man to natural cycles of which he is very much a part. His role, however, is less passive than at the gathering stage. He is so much an active partner in the ecosystem that he modifies it to some extent. In fact, rather a wide gamut of operations is involved. The Eskimo is no less resourceful than the Masaï and is possibly labouring against greater odds, but he harvests natural products that owe nothing to him in their rate of productivity and their local concentration. The Cro-Magnon killers of mammoths showed themselves similarly ingenious, but not "productive" in the sense that some of the African hunters were and are.

The second revolution, *domestication*, opens **Phase B**, the *nomadic–pastoral*. In fact, **Stage 3** (*herding*) is the first tampering with a biological process and thereby is well beyond the mechanical advance that permitted hunting and fishing. The mustering of homogeneous (usually monospecific) populations of animals implies many developing features: territorial control, protection, selection, exploitation, and migration are the most important.

The herders' requirements and influences are therefore many. Their taxonomy is relatively simple as their choice of domesticable animals is restricted, and their choice of grazing and/or browsing plants to which the herd must be led is equally restricted, even if transhumance is involved. Their ecology is certainly more complex and their focus shifts from the autecological or ethological fitness of the wanted animal species (so crucial to the hunter) to the yield of the ecosystem which begins to be perceived as such. Management, for them, leads up from mustering to controlling and from segregating to breeding, and from there to feeding and watering, protection of young, and disease prevention; but it also implies conservation of pasture or browse, fire for prompt (if damageable) sprouting of grass, and eventually fertilizing. Neolithic shelter in caves or in trees is replaced by the Kirghiz tent, the alpine chalet, and eventually the Texas ranch-house.

Psychology of utilization is of paramount importance: on the one hand the taboos against consumption of the flesh of certain animals (cattle, dogs, horses, monkeys, rats, snakes) and on the other hand their desirability for other purposes (traction, fertilization, fuel, social status, religious symbolism).

The third revolution (*cultivation*) opens **Phase C,** *settlement.* Indeed, **Stage 4** (*agriculture*) cannot be taken unless a new land-use system is applied and this has a social prerequisite, namely settlement. Agriculture rests upon a widely improved knowledge, as compared to the previous stages, and it introduces a greater perturbation and transformation of the environment. Agricultural taxonomy is very far-reaching, the "varieties" of some species of plants have as great a morphological diversification as certain natural genera. The ecology of cultivation shows all degrees of insertion into natural stages in the regional dynamics of communities and it offers unprecedented parallels with these same communities, but it is dominated by exotics and domesticates and sometimes realizes wholly artificial equilibria. The channelling of resources runs from mere liberation from competition to fertilizing and to the adduction of a resource not normally provided by the ambient climate; and the transformation of landscape is sometimes seemingly irreversible.

The fourth, or *industrial*, revolution marks the beginning of **Phase D.** It is, in a sense, a full convergence of the first and third, combining in artisanal production the heightened technical skills of tool-making and giving them the social support of a settled society. Such is the primary infrastructure of **Stage**

5 (*industry*). Human economy undergoes one of its major shifts when the ecosystem inhabited by a given group no longer produces any appreciable amount of staple food materials and is therefore geared to processing and not to harvest.

The kind of labour involved, the processing, and the ultimate use of resources are not necessarily different in the instance of a product of mineral and of biological origin: copra, guano, ivory, ambergris, are comparable to petroleum, lime, gold, or sand. However, in an ecological perspective the regeneration, replacement, or substitution in the landscape are at the very centre of our quest. It will therefore be useful to adopt a scale, once more, that runs from the non-renewable mineral to the rapidly regenerating biological source of industrial raw material. We shall thus be following the normal flow of energy through minerotrophic, phytotrophic, and zootrophic relays, as outlined above.

The industrial revolution, as we commonly define it (starting sometime in the eighteenth century), is the fruition of many earlier efforts and achievements. It was contingent not only on technical progress, but of course on demographic shifts of the greatest magnitude. Barbara Ward (1961) has drawn for us the picture of this emergence and the current consequences thereof. In my present outlook, however, it is not so much the weight of the process that is being considered as the nature of the innovation and its ecological impact.

As in herding and agriculture that have so often added their powers, it is inescapable that **Stage 5** (*industry*) and **Stage 6** (*urbanization*) be closely parallel. The gamut of clan, village, town, city, metropolis, and megalopolis shows a dizzying spiral of energy concentration. Each order of magnitude having its constraints and requirements. The modern city building would seem to present the utmost development in freeing the individual from immediate environmental strictures: light, heat, water, food, entertainment, commercial, intellectual, and religious needs can all be satisfied within a limited space and by elements that cannot originate locally.

Insulating himself from the natural ambience, urbanized man bears with or selects the objects and activities that make his real world. This involves drawing from very vast resources, in fact, from the whole world. In an executive office, the structure of the building will be iron (mined in Schefferville, Quebec, processed and cast in Pittsburgh, Pennsylvania), cement from Vermont limestone quarries, sand from Lake Champlain beaches, marble from Peru. The office furnishings will involve nickel from Sudbury, Ontario, tin from Bolivia, redwood from California, mahogany from Costa Rica, ivory from the Congo, gold from South Africa, and plastics from New Jersey. The carpeting from Iran and Tibet is made of sheep and yak wool, the leather upholstery is Spanish, the wrought-iron is Italian, the cushions are of Japanese silk. A modern French painting and an Eskimo print are on the wall; the desk is laden with papers from the mills of many countries; glass and ceramic pieces are Venetian, Portuguese, and Puerto Rican. Et cetera, and again et cetera. Looking back from these objects at their present location and through the transport and processing and all the way to the origin of the raw materials, one gets a glimpse of the many ecosystems that have produced (and lost) them: the redwood forest, the nickel mine, the herd of yaks, the parasitic silkworms; the sawmill, the deep-reaching shaft, the rows of mulberry trees; the sweating lumberjacks and millhands and railroadmen, the grimy miners, the furnacemen, the hardware dealers, the busy growers, the patient pickers, the spinners and weavers, and the astute shopkeepers. This great centripetal movement involves a tremendous flow of energy from the geological and biological processes of resource accumulation through the many transformations accomplished by photosynthesis, storage, growth, differentiation, and decay. Thus, a niche (the executive's office) in an enormously complex ecosystem (Montreal) is seen to draw from the greatest possible variety of sources.

In return, information flows out of this niche, and through

several relays increases or decreases or otherwise controls the redwood forest, the yak herd, the mulberry grove, the sweat-shop, the mine, the ivory tower. It is easy to predict the eco-logical transformations that employment–unemployment, in-vestment–withholding, and other economic and political al-ternatives will induce. Wars and invasions have effected many such changes, which were triggered at the focal points where command of information is concentrated.

The post-industrial period in which we now live has been analyzed and characterized so well by Kenneth Boulding (1964), Lewis Mumford (1966), John Kenneth Galbraith (1967), and its spatial future so sharply outlined by Con-stantinos Doxiadis (1968), that it is tempting to re-draw its texture from their canvasses. I must be content, in the present context, with acknowledging my indebtedness and with a brief basking in the light they have cast on our world and on ourselves within it.

A better methodology for the measurement of man's habi-tat is very much needed. The kind of inventory which I have sketched into my descriptions of cities is made in a strict parallel with animal habitats and according to uniform eco-systematic criteria. But it calls for a careful transposition of economic, social, and historical data and processes if we are to develop a true science of human ecology. This is an urgent task for we are acceding to new dimensions of environment.

The fifth revolution (*cybernetic*) is at the origin of **Phase E**, the *climatic-cosmic*. It extends the gamut of communica-tion so far beyond the signalling systems and the symbol-transmitting media that the ecological area of participation is considerably increased. McLuhan's (1964) magic formula, "the medium is the message", appropriately unveils a modern realization of the inseparableness of form and content. The cybernetic revolution is borne on the infrastructure of a tech-nology that is always referred to as "sophisticated", since the workings of television and of computerizing machines and of satellites only became possible in the light of our ultimate advances in physics. It is, however, an essentially psycho-social revolution in that it strikes in a new fashion at the perception by individuals of what is knowable and imaginable at the same time that it provides virtually every man with unprecedented means of extending himself.

Stage 7 (*climatic control*) brings us well beyond the powers of mining, agriculture, manufacturing, and town-building. The seeding of clouds to displace rainfall patterns, the explosion of atomic and hydrogen bombs, are truly phenomena of geological magnitude, equal in power to glaciation and volcanic eruption.

Stage 8 (*exobiological outburst*) marks an even greater breakthrough. "Reaching for the moon" has been pulled right out of mythical imagery and reset as scientific fact. The place of such knowledge in the conscience of the individual is, so far, very insecure. The newspapers recently reported an interview with a man known to be 130 years old, born a slave in the Southern U.S.A., and happy in his illiteracy; he denied the truth of so-called news to the effect that men had been walking on the moon. More educated people do not believe it either, in the sense that they have not allowed it to make any difference in their inner world. The greater recognition is yet to come.

Thus these five revolutions stand as beacons in the development of man's dominion over environment. They can only be seen as cumulative. From the axe to the computer, technological inventiveness enlarges its focus from single objects (gravel, tree, bird) to unseen forces (wind, gravity) that are known, not experienced, calculated, not touched. The implicit science of early domestications is eventually explained by the unravelling of the genetic code and simulation techniques are applied to breeding programmes. The purposeful modification of the landscape itself, initiated by neolithic fire-setting and more deliberately patterned by sowing and planting, moves from the closed economy of subsistence to specialized production of response to industrial demand and long-distance supply. The impact of industry and urbanization, in their current phase, results in new distributions of the wild–rural–suburban–urban occupancy. The resource base of land-use patterns is totally altered by changes in the communication network that transmits material goods and information with great immediacy. Thus, in the end, the interaction of knowledge, need, purpose, and patterning has undergone a displacement of relative weight in decision-making that calls for a more rational and more pervasive planning.

DESIGN AND PLANNING   V

Does the snow-goose, airlifting itself from the warming shores
of Chincoteague Island, visualize the Arctic tundra towards
which it is heading to breed? Does the bee, radar-beamed to
a particular tree in an apple orchard, have a conscious dream
of yellow nectar in the unfolding white blossoms? Does the
beaver, felling his first trees across a narrow stream, have a
mental image of the pond that will rise behind his dam?

In other words, what are the biological origins of environmental design? This term must successively be understood in
its first meaning as *intention* and in its second acceptation
as *plan*. First comes the wish, and then the project. The wiles
of hunters were directed to a prey that had already proven
satisfactory. In fact, calculated risks were taken in which the
desirability of various kinds of game was weighted against
their accessibility. Further considerations involved prediction
of its periodicity or renewability and cost of labour in securing
it. Snares, pits, fire, killing instruments were variously devised
only to become gradually obsolete with the advent of domestication.

Although implicit designs can be read in the beaver dam,
and somewhat more explicitly in the hunting and herding
human communities, it does not seem very appropriate to
trace planning back any earlier than the settlement phase and
the onset of agricultural activity.

The kind of foreknowledge that agriculturists have shown

concerned not only the life-cycle of the plant and animal but also the qualities of the site that would or would not sustain a periodic productivity. Thus the various groups of men who created seeding and planting economies carried some tested plants with them and sought to fit them into new ecological niches. The development by American Indians of a very large number of local races of maize demonstrates the rich adaptability of this cereal and also the intelligence of the breeders.

What is the initial point of land-management? The setting of fire by hunting tribes to flush out game? The clearing of forest by shifting settlers? The building of terraces by the Southeast Asians and the Mediterraneans? The introduction of rotation, first by swidden farming and later by peasant farmers?

As we look back upon these accomplishments, we marvel at the discovery of the axe, the wheel, and the plough, the use of fertilizer, and the devices for storage. It is not always easy to probe the knowledge which must have been preliminary to the development of these ever more powerful means of dealing with the resources of the environment. We are bound to work backwards from our own kind of awareness of soil, plants, and animals, and to classify in our own way the diverse land-use patterns that confront us today. Most fortunately, we are still able to witness landscapes where gathering has not been entirely swamped by agriculture, where hunting and fishing, although more recreational than sustaining, still impose their requirements.

Taking stock of our resources on a planet-wide basis, and devising uniform categories that will allow much-needed comparisons, are a relatively recent endeavour. In its fully explicit form, it merely dates from 1949 when the Commission on World Land Use Survey was formed by the International Geographical Union. Its first report was presented in 1952.

How did geographers then perceive a workable taxonomy of land-use types? They proposed a framework of nine major divisions within which all existing management regimes could be included: 1. Settlements and associated non-agricultural lands. 2. Horticulture. 3. Tree and other perennial crops. 4. Cropland. 5. Improved permanent pasture. 6. Unimproved grazing land. 7. Woodlands. 8. Swamps and marshes. 9. Unproductive land. It goes without saying that each of the major categories can be, and has been, subdivided for application to tropical and temperate areas and also to meet the requirements of different scales.

It is easy to deduce that the constraints of mapping were foremost in the minds of the originators of this world scheme. The late and beloved Dudley Stamp was the driving force in this initiative. His own admirable analysis of the British landscape and rigorous standard of land-use mapping for the United Kingdom stand as the prototypes for all such endeavours.

Many monographs accompanied by maps have now been published. The Canada Land Inventory stands out as one of the major contributors to this relatively new discipline. It is based upon a somewhat simplified scheme that comprises six categories: I. Urban. II. Agricultural lands. III. Woodland. IV. Wetland. V. Unproductive land. VI. Water. The application of this system to quadrangles of the Canadian territory at the approximate scale of one mile to one inch may number some twelve to sixteen subcategories.

It would be most rewarding to analyze the adaptations of the World Land Use Survey in countries with such widely different attitudes to man and to nature as Japan, the Netherlands, the U.S.S.R., France, Brazil, New Zealand, Britain, and Canada. One would surely witness differences in the relative weight of analysis of natural vegetation, rural exploitation, mining, and industry and, of course, urban development. The accelerated reversal, in this century, of the urban/rural balance, the accepted fact that the city controls the country, and the corresponding increase of urban studies have completely redirected our perspective on land-use. The socialization of man, of his education systems, and the influx of technological and scientific values find their achievement in urban living. The true confrontation of man and environment now takes place in the city. From the executive tower, the conference room, and the picket line, urban man actually dominates and controls the whole range of the landscapes through the suburban sprawl and the farm land to the distant forest and stream.

Re-patterning by man has taken on many aspects. Since we are looking at land with ecological binoculars, it is appropriate to consider first an acceptable taxonomy that is consonant with the principles already outlined and then to probe as deeply as we can into the functional aspects that underlie a consistent land-use management based upon its true potentials.

In order to achieve this, it is useful to resort to the same rationale that underlies the evaluation of man's power to transform the environment. The processes that characterize man's adjustment to a system such as ploughing-sowing-harvesting-storing in agriculture are the elements that cast their stamp upon the land and that have to be assessed in order to convey ecological characterization. And thus some landscapes turn out to be exclusively agricultural or industrial, or urban. In a good many, however, a complex mosaic prevails where the inter-system exchange, such as the marketing of grain, is more important than the intra-system cycling, which runs from sowing to harvesting.

Picking up the ecological theme as it was initially proffered in the first chapter, we find that some landscapes are overwhelmingly dominated by mineral resources (the desert, the tundra), by plant-life (the rainforest, the grassland), or by animal life (the seal-islands, the bird-cliffs). This dominant feature of the natural background may be radically changed if the desert area becomes an industrial oilfield, if the rainforest becomes a rubber plantation, if the seal-islands become a military base. Such applications of man's power modify or even reverse the relative activity of the six trophic levels as defined in the previous chapter.

From the point of view, therefore, of ecological dynamics, the scale of land-use patterns runs from wild to urban, through rural, industrial, and suburban types.

The wild landscape remains as the basic unit for constant reference to abiding environmental forces. Likewise, no doubt, the wild man, animated by wild thought, is our guide to elemental perceptions. Lévi-Strauss' "pensée sauvage" (1962) is at the root of physiological gratifications and harmony in the sharing processes, and offers its infrastructure to environmental consciousness.

The scientific arsenal which has been geared to the study of wild (or so-called "natural") landscapes has been inventoried earlier on. I drew attention, for instance, to the vast compilation of information on primeval environments that will result from the efforts of the International Biological Programme. Our knowledge, therefore, at least our specifically ecological knowledge, of wild environments is much better than our knowledge of man-influenced or man-made environments, since ecology developed in the fold of the natural sciences and not of the sciences of man.

This is neither fortuitous nor regrettable, for I firmly believe that all study of rural or urban ecosystems must seek its origins in the wild. The principal reason is this: in wild ecosystems the cycling of resources bears a time-relationship to evolutionary and ecological processes which is not carried over to rural or urban systems.

It is worth spelling this out at greater length. Every species of plant and animal, indeed every variety (including the races of man), has a limited genetic potential within which are inscribed characteristic requirements, tolerances, and capacities of using resources. Thus the sugar maple and the red maple, sometimes growing together but mostly on different sites, do not fit into the same soil–moisture gradient, for the red maple will occupy both wetter and drier sites. On the other hand, given a well-drained upland, the sugar maple will compete successfully with beech, linden, ash, and other trees, whereas the red maple will soon be eliminated altogether. If we turn to an old field in Georgia, we find it successively occupied by grassland, savana, pine forest, and oak-hickory forest, and an accompanying replacement of breeding birds is to be witnessed, such as the grasshopper sparrow, the field sparrow, the pine warbler, and the red-eyed vireo. A complete run-down of all plant and animal species found at any one time and place to share an ecosystem allows us to chart their overlapping *ecological amplitudes.*

Much as I should like to dwell upon this aspect of environmental adjustment and to quote numerous well-documented examples, I must be content to draw attention to the fact that the resolution of internal and exernal conflicts in the adaptation of an organism to one or more ecosystems implies two series of phenomena. The first is the ability to cope with the opportunities and adversities offered by the resources of the ecosystem. The second is to achieve a relatively secure status in the sharing order. In other words, the sugar maple and the red-eyed vireo are both well adjusted on the one hand to the amount of light, heat, water, and food made available to them in the forest, but each has also found a proper *niche* where their co-operation/competition capacity is in a state of long-term equilibrium with those of beech, ash, linden, and other trees, and indeed with the shrubs and herbs in the example of the maple tree; and with other breeding birds (whether prey or predators) and insects or mammals.

The wild ecosystem, therefore, is seen as the result of almost endless trial-and-error in the heredity/environment contest. Evelyn Hutchinson (1965) has well described "the ecological theater and the evolutionary play", and, in a sense, every ecosystem can be seen as an act, never the final one, in this drama.

This amounts to saying that, in the wild ecosystem, mutual adjustment through actual genetic selection has resulted in enduring patterns of resource-sharing. A built-in mechanism of interlocking cycles seems to regulate the system, and im-

72

pacts from other ecosystems are either absent or have been constant in kind and in strength for a long period of time. Such would be an alpine meadow, a bog, a dune, a maple forest, a bird cliff, etc.

Such are not an agricultural plot, an industrial park, a suburban development, or a city. In all of these, it could be exaggerated to say that no mutual adjustment in time has taken place, since many cultivated plants as well as weeds and animal pests show considerable adjustment to man-made habitats. But the adaptation is primarily ecological, not genetic; the coexistence is planned, not spontaneous; made stable by repeated deliberate, intelligent, and experimental interference.

The revolutions operated by man in his ever-growing power to redirect the forces of nature have been outlined in the previous chapter. The agricultural turnover and its wide-ranging effects are at the origin of the fundamental discontinuity of processes that we must witness between the wild ecosystems and all the others.

The impact of the noösphere, as Teilhard de Chardin and Vernadsky have called it, is well detected by Lévi-Strauss (1962), in the following quotation:

To transform a messy herb into a cultivated plant, a wild beast into a domesticated animal, to draw out in either of them nutritional or technological properties that were originally completely absent or could barely be suspected; to make solid and impermeable pottery with an unstable clay, liable to crumble, pulverize or split (but only after having determined, among a multitude of organic and inorganic substances, that which is best fit to cleanse, as well as the appropriate combustible, temperature and duration of cooking, the efficient degree of oxydation); to elaborate the techniques, often time-consuming and complex, allowing to grow without soil or without water, to transmute toxic seed or roots into food, or else to use such toxicity for hunting, war, ritual, must have required, we hardly can doubt it, a truly scientific mental attitude, an assiduous curiosity in perpetual awareness, an appetite for knowledge for its own sake, since a small fraction only of the observations and experiences (that we must suppose to have been inspired first and foremost by the craving for knowledge) could give practical and immediately useful results.

This somewhat breathless statement, better read than spoken, is more in the style of Marcel Proust than Claude Bernard. It is nevertheless in the main line of thought initiated by both the experimental and the introspective schools.

With what physical and mental equipment does agricultural man work himself into the established fabric of wild ecosystems? Ecologically speaking, what is a farm? What goes on at all six trophic levels?

The sunshine and rain, the radiant heat and energy, are unchanged, whether the ploughed acres were formerly a forest or a marsh. But the soil itself is the result of at least hundreds, often thousands, of years of weathering and of exploitation by wave upon wave of plant and animal masses. Its mode of productivity, whether high or low, is best suited to a certain kind of uptake—by trees, for instance on a rocky moraine, by herbs in a fine-grained periodically inundated flatland. The latter, in the St. Lawrence and Great Lakes lowlands, is widely reclaimed by the farmer, who stops or drastically reduces the seasonal flooding, thereby making the muck and its underlying clay more amenable to cultural practices. This brings to a stop the age-old soil-forming processes: yearly increments of silt and alternation of organic decomposition under water and in the presence of free air.

Native plant-cover has been stripped off from the beginning (elm trees, cinnamon ferns, Joe-pye weeds, touch-me-nots, and the whole cohort of diversely rooted and expanding plants). In their place, an exotic from Mesopotamia (wheat), or from coastal Europe (beets), or from the Andes (potatoes) has been substituted. Its efficiency in producing grain, root, or tuber, protein, starch, and sugar is dependent upon freedom from competition on the part of spontaneous invaders that pour in from all around. The incapacity of any one species of plant to tap all the resources of the air and soil leaves so much unsaturated material that every cultivated field is a vortex that sucks in the hungry weeds. The resulting growth is eventually harvested and for the most part taken away, that is, out of the ecosystem altogether and sent to the market. But some of it may well be ploughed back into the soil, thereby restored to the mineral level, whereas some is immediately or eventually consumed at the next higher level by fowl, pigs, cattle, horses. A toll is taken by pests, of course, many of which are animals, especially insects, although fungus parasites consume plant materials too.

At the level of investments, an old-fashioned farm, for instance in Quebec at the turn of the century, showed remarkable autarky (or self-sufficiency). Water was drawn from a well. The houses and their dependencies were built with stone, gravel, and sand of local extraction and/or with wood from the woodlot which also served as fuel. Furniture and household appliances were likewise hewed out of indigenous resources. Staples like sugar from maple trees, eggs, butter, milk, and meat were produced on the farm. Wool and leather were available for clothing, and even flax for linen. The relatively large surplus of primary (vegetable) and secondary (animal) products, stored and eventually exported and sold by the farmer, permitted an import that would compensate for the depletions at all levels: fertilizer for the soil, seed for the crops, and locally unproduceable vegetable materials for man (tea, coffee, spices) or beast (molasses, salt, fishmeal). Moreover, in this exchange, necessities were satisfied and amenities were offered in the form of glass, nails, china, books, clothes, etc.

The farmer's control, in the last analysis, may well be geared to an ingrained tradition that is so much a part of himself as to operate as an inner compulsion: rotational practices, form and function of implements, routine of the calendar year, are all so fixed and so unquestioned in Thomas Hardy or Emile

Zola's country-folk that they may well fit into the self-contained feedbacks of the ecosystem. However, information from other ecosystems (especially the city) eventually reaches even such a seemingly closed world: new agricultural practices, new varieties of plant and livestock, opening and narrowing of markets, legislation concerning pests or credit, all have the power to modify the cycling of mineral, plant, and animal materials on the farm.

The power of the farmer has to exercise itself constantly over his acres. There is an optimum state of soil in each field for the intended crop; there is a necessary sequence of practices for maximum production and sustained yield; there is an optimum number of grazers and feeders; and there must be constant defence against invaders, against the return of primeval and other spontaneous plants and animals. In other words, the secondary equilibrium established by the farmer is not capable of maintaining itself without repeated intervention. It is under constant menace of an offensive return.

There are so many different kinds of agricultural landscapes that they can hardly all be superimposed on the picture painted above. The development of monocultures and the industrialization of agriculture have brought about countless new patterns. For instance, an orchard or an extensive cereal field or sugarcane plantation are totally geared to primary production and animal life is utterly stamped out as far as is materially possible. The ecological standards that I am applying to landscape are precisely devised to bring out these functional differences. Thus the relative weight and strength of cycling activity at each trophic level and its relative dependence upon outside influx is the principal feature that will distinguish one ecosystem from another. In this light, a traditional Quebec dairy farm with its five- to seven-year rotation, a Filipino rice-paddy with its fish-culture, a Brazilian rubber plantation, and a New Zealand sheep range stand in very strong contrast in all respects.

Industrial developments are predicated upon functional

premises that are bound to relate to environment in a very different way from rural economies. The relation of an industry to its resource base is normally sought by an ecologist by looking for the source of its raw materials. The existence of a textile industry in areas that grow no cotton or of an aluminium plant far from bauxite deposits makes us shift our attention to power and to labour. Whether docile labour or cheap power bore the ultimate weight of decision, there can be little doubt that the policy of industrial location was dictated by economic considerations. The ecological consequences, in Western Europe for instance, have been amply documented: urban slums, industrial pollution are graphically recorded by Charles Dickens, Emile Verhaeren, Eugène Sue, and the low quality of human life is stressed. The growing consequences on the environment as a whole did not result in the emergence of a conscious requirement for planning. Until quite recently.

The ecological impact of the industrial revolution, compared to those of the agricultural revolution and placed upon the matrix of the six trophic levels of ecosystem dynamics, gives us a new reading of man's exploitation of environmental resources. Taking the simple example of a textile industry in an Eastern North American small town, as a parallel to that of an early twentieth-century farm, we can scale the ecosystem from mineral to control level.

Mineral resources are provided in the form of cheap hydro-electric power generated by the dammed rivers not very far away. Construction materials for the factory may well be drawn from nearby quarries and pits. The substratum is otherwise completely sealed off by roads, yards, and buildings, and in no way capable of sustaining plant life, save for sporadic weeds.

This being the situation, vegetable materials are entirely imported, principally fibre from cotton grown in the Gulf of Mexico or the West Indies.

All animal life is as drastically excluded as possible, except for the human animal whose metabolic processes are harnessed to mill operations.

Heavy investments are made in machinery imported from much larger industrial or industrial/urban centres. Furniture and furnishings, office and factory materials of all kinds are drawn from many other ecosystems. The local population provides the bulk of the labour, but the managerial force is more usually unrelated to the town historically and biologically. The capital upon which the enterprise functions is also provided from a distant source.

Control is immediately in the hands of the managerial unit, but ultimately in the grasp of absentee capitalists.

Such an ecosystem, therefore, has a very small measure of autonomy. The landscape which it occupies provides a substratum for building, a river for effluents, and of course light and heat from the atmosphere, but fuel and power have to be imported. The native plants and animal life have been completely destroyed and not replaced save by a few weeds and houseflies. However, plant and animal stuff is imported as food for humans. The ultimate product is exported as yards of cloth or in the form of further processed merchantable material. A variable percentage of the wholesale revenue is reinvested in the factory as salary and maintenance and the rest remains outside the ecosystem as dividends to non-participants in the production.

Again, as with the agricultural ecosystems, many variations are known, according to the size of the enterprise and its rural vs. urban location and, above all, its relation to regional resource bases. Thus, in France, Spain, or Italy, an olive oil or a wine factory will have spatial, economic, and human continuity from the mineral basis through the vegetal production to human investment and marketing. But the steel industry of the Great Lakes or of Japan, the shipbuilding industry of Scotland have no such relation to the landscape in which they operate.

Urbanization in many ways permits the ultimate emancipation of man from his environment. In the first place, from climate itself. The tent of the Kirghiz and the wigwam of the Algonkin buffered the variations of temperature and kept out the rain, but the modern city building filters in the masses of air and maintains them at an even temperature and humidity; power drawn from hundreds of miles away provides heat and light. The buildings and pavings are all made from materials extracted from quarries or mines and variously processed industrially. Although the mineral resources of the urban site are sealed off, as it were, the mineral forces do not stop their play. The erosion of buildings, the frost-heaving of pavements, the destruction by wind, the swelling by hot moisture, all claim a constant defence and repair. It remains that the intake of mineral resources, including energy, far exceeds the load of the intra-system contribution in the city.

At the plant and animal levels, this is even more so: fibre, wood, grain, vegetables, milk, eggs, meat, skins, in various forms, natural and processed, are imported in massive quantities to build, furnish, clothe, and feed the human population.

What plants are grown (grass, flowers, trees) and what animals are raised (goldfish, canaries, cats, dogs) are in no sense productive, but figure at the level of amenity investments, responding to a psycho-social not a physiological need. This distinction, already of some importance in much less modified environments (witness the very early taming of the dog as a pet), becomes more weighty in the urban environment, where the range of physiological/psychological requirements is much widened. A large, ungrazed, unused lawn is a status symbol, not a source of food or a provider of health; a beautiful Afghan hound is likewise a sign of elegance and not a retriever of prey.

The dependence of the urban ecosystem upon other ecosystems for a massive inflow of mineral, vegetable, and animal matter, and the consequent inactivity of the ecosystem at its four lowermost levels, are compensated by the extent of its investments and the strength of its power to control many other ecosystems. Any minor or major catastrophe (a fire or a hurricane) draws attention to the wealth of cities, to the heaviness of invested stone, wood, metals, manufactured goods, works of art, etc. "Damages" quickly run into the millions as we are used to calibrate such values in our society. A fraction of this would account for the total productivity of an equivalent area in a wild ecosystem such as a forest, or a rural ecosystem such as a dairy farm.

But the most heavily laden trophic level is certainly that of control. The tribal chieftain, the neolithic artist, the rural schoolteacher, the parish priest, and the local professional certainly dispensed information that had an impact on the uptake of water, the discriminating plantation and breeding habits, the construction of shelter and storage of food, the social structure and habits, and the inter-system relationship. But the modern city, especially the megalopolis (and tomorrow the ecumenopolis or world-city), is such a pervasive force as to have become very visibly responsible for the management of virtually all other ecosystems. International corporations have a tight hold on fuel and power, are extending their empire to communications and hostelry; governmental takeovers and intergovernmental agreements regulate mineral exploitation, agricultural production, wildlife protection, and variously tax and subsidize industrial production and commercial distribution. From the executive tower and the civil servant's desk comes a message that stops quarrying, that forces wheat land to remain fallow, that halts salmon fishing, that taxes building materials, that prohibits public gatherings, that legalizes drugs. These are all pressures applied upon the whole gamut of ecosystems and at each of the trophic levels: mineral, vegetal, animal, investment, and of course control itself which is always contested.

It will not be possible to fully explore the variety of landscapes that fall within these principal ecological categories: the wild, the rural, the industrial, the urban show up not only under many aspects but in various modes of compenetration. I hope we can have set our sights in a truly ecological perspective by leading up to this classification through an initial exploration of human perception through the scientific justifications it has achieved, and through the detection of the revolutionary processes that have variously increased and possibly improved man's ability to cope with his environment. With these points of reference in view, we can turn to management itself and examine the criteria implicitly and explicitly accepted by analysts and planners for the determination of environmental potentials and the projection of their designs. This cannot be done without reference to acceptable and unacceptable social goals. The inscape/landscape pattern is more than ever subject to ethical ponderation.

We had to look back from the moon, before we fully realized that we have "only one Earth". The satellite pictures confirm many of our painfully achieved mapping efforts, but they also give us a greater feeling of authenticity, since it is now physically possible to see whole continents at a glance. The power of dreams is still with us, however, not only to devise new means of reaching farther out into space but in the interpretation of the "sea around us", the "restless atmosphere", and the earth between our toes. For instance, we do not directly see how our planet is at once sparsely populated, with immense areas of wild landscape, and overstressed by an alarmingly high rate of human multiplication and a pressing demand for raw materials, industrial products, and cultural amenities. A sharp man's-eye view detects those starlike nerve centres that are the megalopolitan developments, and the increasingly tight and solid neuron filaments that link them in a worldwide network. Reading the worldscape is too new a sensorial experience not to be heavily polarized by non-visual knowledge. The observers are bound to wear differently tinted glasses. To some, the abiding beauty of mountain peaks and ocean abysses will send reassuring messages; to others the cratering of Viet Nam and the devastation of the badlands will stand out; and to yet others it will be what Philippe Sollers has called "le cauchemar panoramique des états-munis".

The ecologist, wedded to his criteria of the sharing of resources by diverse exploiters and to his search for communication and energy transfers, is hard put to "stick to his guns". I hope, however, that I have been able to demonstrate earlier in what sense psycho-social forces, and of course perception itself, are in the position of levers in the ecosystem, since they so often activate or inhibit a particular resource circuit, whether water, plant or animal species, building material or behaviour pattern.

What then can ecology contribute to ethics, private and public? And how can ecological law be recast into an ethic pronouncement? It cannot. But it can provide a much improved background to a real and realistic assessment of human and social responsibility. As the developing science of psychology has changed (often reversed) the consensus of certain societies concerning the rightness, or at least the permissibility, of certain behavioural patterns, so an improved knowledge of ethology has led to a less delusive ethic. Emerging moral judgments displace the weight of guilt and of sanction in such a way as to modify the prevailing stresses in a given society. This amounts to an ever-shifting definition of responsibility as science defines ever more narrowly the actualities of man's control of himself and his relation to weather, vegetation, animals, investment, and information.

Maybe the identification of these stresses as they are revealed to us by the study of a large gamut of ecosystems in which man is a partner, will allow us to consider the nature of the benefits and the acceptability of the justifications. If some rights of man to a share in environmental products (oxygen, food, shelter, money, travel) are more fundamental than others, on what grounds may this be argued? And it will need to be considered: at what expense and to whose (critical) detriment at given times and places? Admitted that the right to breathe is more imperative than the right to own an automobile, what hierarchy of rights to draw benefits from the environment could we recognize?

A truly ecological approach to rights-and-duties can be attempted by referring to the fundamental dynamics of the ecosystem as outlined above. Aldo Leopold, in 1949, tried to formulate a "land ethic", and in many of their writings Paul Sears and Frank Fraser Darling developed this theme with an ever-sharpening urgency. Howard T. Odum (1971), clearly following this trend, and no doubt much influenced by his sociologist father, goes so far as to propose "ten commandments of the energy ethic for survival of man in nature". For my part, in a recent essay (1971) I picked up these threads and in a graphic outline of the ecological basis of human rights I tried to place human necessities and human predicaments within my framework of six trophic levels in search of a valid reference for responsibility and eventual legislation. The products at each level are in some way indispensable to the proper functioning of human ecosystems, but for widely different reasons, and in widely different constellations at different times and places.

As I see them, the *benefits* are minerals, vegetables, animals, functions, services, and information. The *justifications* are physiological, psychological, social, economic, political, and religious. The inter-relatedness of the benefits and justifications can be tested by comparing the facts and values yielded by ecology, ethology, economics, ethnology, ethics.

The anatomy and physiology of a landscape and of its several ecosystems are strongly marked by the interaction of the five E's, at all six trophic levels. I should like to run the whole gamut of man-and-landscape to probe the relative weight of these five major components, but I must be content with one example and will take it nearest to home, in French Canada.

1. Ecology. The St. Lawrence Valley was colonized in the seventeenth and eighteenth centuries by people mostly from Northwestern France. It comprises cold-continental, alluvial lowlands and two hilly and mountainous districts. The original plant-cover was almost entirely forest, essentially of two kinds: broadleaved deciduous hardwoods and needle-leaved evergreen softwoods, interspersed with wide rivers and abundant marshland. The fauna was abundant and diversified, comprising hordes of forest ungulates, flocks of migrating birds, a great variety of fish, and a rich insect fauna.

2. *Ethology.* The human population had to face long cold winters. Indians used native plants and animals for food, wood and skins for shelter and clothing, and so did the French who also imported plant and animal products. In the two groups, mating customs and rearing of young differ markedly, as do the whole social organization and the relay of commands.

3. *Economics.* Early management of the landscape was geared to very little gathering but a good deal of hunting (massive export of pelts) and a growing agriculture, later complemented by marketing and urbanization; and more recently by industrialization and metropolitan growth. Export/import and capital investments underwent drastic changes of polarity, from Indian nomadism to French colonial dependence, to British colonial tribute, to U.S. empire.

4. *Ethnology.* French Canadians were racially rather homogeneous; limited cross-breeding with Indians did take place in early times, and later on intermarriage with the British was widespread and more recently with many other ethnic groups. The rural large-family-centred economy gradually yielded to the urban small-family unit. Church and school were focal community points, almost exclusively so for some time. Educational practice favoured an élitist cultural development with very low scientific and technical content.

5. *Ethics.* Until rather recently the power of the Catholic

Church to impose its prescriptions was virtually unchallenged. Western European bourgeois formalism was the vehicle of this rather strict code of behaviour. Its somewhat masochistic and other-worldly orientation resulted in a low grasp of environmental opportunity and a lack of initiative in tapping natural resources.

The forested territory of the Laurentian Shield was unprofitably turned into farmland under some economic pressure, but mostly in response to a crusading (ethical) spirit. The resistance to modernization of agriculture amounted to a myth (Brunet 1958) which entailed not only enduring political consequences (such as predominance of rural over urban counties in Parliament) but, above all, rejection of new technology. At the same time, lack of local opportunity forced migration to nearby heavily industrialized New England.

The ethical constraints that favoured (indeed enforced) large families tended to perpetuate a cheap labour market. For instance, the relative productivity and profitability of the maple-sugar industry (Dansereau 1945) was derived from a very appropriate timing (in the slack between winter and summer activity) of the large-family occupation-pattern. Other yields and pressures upon the resource-stock comprise the rejection of mutton as food by French Canadians, whereas the low-grade pastures of both Appalachian and Laurentian hills would have been profitably exploited by sheep, and the saltmarshes (as in France) might even have yielded a high-grade product ("agneau de pré-salé"). Falardeau (1967) has carefully scanned the French-Canadian novel for its implicit and explicit socio-economic themes and found there much evidence not contained in the scientific and explicitly sociological literatures.

The ecological content of existing practice and legislation in Canada and elsewhere is not too apparent. We can only scan the books and search for the intentions of the legislator and beyond that for the pressures of vested interests and those of public opinion in order to inform ourselves of the perceptions of environment that may have been involved.

The history of the scientific impact of ecologists upon technology and upon legislation, in fact, follows a pathway, from 1872 to 1972, roughly parallel with the scale of environments I have proposed in the previous chapter. That is to say that the nineteenth and early twentieth century conservationists were almost exclusively preoccupied with saving wild plants and animals. National and regional parks were "set aside" because it was evident that a minimal "Lebensraum" was essential to the survival of menaced species. These early post-Darwinians had been shocked by the extinction of the dodo, the great auk, and the passenger pigeon, and urged the powers-that-be to take action. Hunting and fishing of both commercial and game animals were increasingly subjected to rules and penalties. Not without stress, need it be said, in the face of more or less thriving industries such as commercial fishing already at the mercy of seasonal variation and fluctuating infestations. Conflicts also arose from unacceptable values, as with elephants and rhinoceros and birds-of-paradise nearly exterminated to suit the fancies of tribal chiefs, of rapacious traders, and of "la dame de chez Maxim".

It is most intriguing to probe the ethical background of conservation legislation and to unravel the mixture of motives that powered the preservation of stream, forest, and wildlife. It was the genuine disinterestedness of scientists and the religious dedication of certain African tribes that ensured complete preservation of certain streams or sacred groves, and thereby the salvation of rare species. It was the aristocratic desire of privacy on the part of Spanish noblemen that saved the lynx from extinction in Southern Europe and maintained the great marshes at the mouth of the Guadalquivir as a "chasse gardée" in a near primeval state. It was wealthy New Yorkers and Bostonians who preserved green belts (more recently become semi-public property) around their megalopolis and who saved Eastern Quebec streams from excessive

87

lumbering and over-fishing. It was the maharajahs who made sure that their grandsons would have their quota of tigers.

Meanwhile, to be sure, the less privileged had access to wild land and to game birds and beasts. Poaching on the one hand and over-protection on the other introduced many upsets and proved the lack of wisdom of unscientific laws. Massive death of deer in Michigan, Wisconsin, and Minnesota for lack of adjustment to the food supply and undue suppression of wolves; the erosion of soil and decimation of plants by overgrazing elephants in the Congo; the halt of regeneration of beech forests in New Zealand because of introduced European deer and North American elk, all point to unwise practices.

What the conservationists discovered, sometime in the Forties, was the uselessness of their eloquence when it confined itself to "saving" and to "setting aside". Clearly a better application of ecological principles to legislation on land-use had to expand on a broader compass. It is with the emergence of the theory of the ecosystem that *resource administration* came to replace conservation, in its traditional sense. Two American ecologists, Paul B. Sears and Stanley A. Cain, were very persuasive spokesmen for this new orientation, especially as they were able to implement it at the level of the university curriculum at Yale and Michigan. Actually, such blossoming on the campus was the result of the convergent winds that had carried seed from the natural sciences and the social sciences respectively. It was also favoured by a greater social and political affirmation on the part of university personnel. (Admittedly this phenomenon was newer in America than in Europe, although on both continents very little decisional power was allotted to academics.)

It can be said that in the early Fifties ecological science was given a new lease and an enlarged scope. University graduates who had started in the natural or the social sciences and who had flocked towards resource-oriented degrees were much in demand. The dimming of the ecological vision in the late Fifties and early Sixties, at least in natural science circles, is hard to understand retrospectively, since population growth, urban sprawl, spoilage by industry and by war had by then reached unprecedented proportions. The breakthroughs in molecular biology, nuclear physics, and information theory may well have cast such blinding light that the scientific establishment (university–industry–government) neglected its earlier investments and put the full weight of its resources in these more newly developed fields.

Now that ecology has become a by-word, almost a password, that its intentions (if not its meaning) no longer have to be explained at cocktail parties, political meetings, and religious gatherings, it would seem to command the fullest possible approbation. The substance and the immediate purposes of such widespread support have yet to be more sharply defined. The change of gear among old-time ecologists, the assumptions of the converts, and the claims of the new generation do not necessarily permit very smooth driving of this new bandwagon.

From protection of rare species to conservation of representative tracts of land, on to the recognition of the need for multiple use, the economic and the ecological forces have repeatedly clashed. But now, the economists and the ecologists have apparently achieved an acceptance of each other's vocabulary, and initiated experiments with each other's concepts and methods, and have thus uncovered some common ground for research. The frame upon which I am at present stringing my observations and developing my arguments lies upon such premises. A six-level model of the ecosytem has turned out to be useful in harnessing a multidisciplinary group to an interdisciplinary endeavour. It will be of the essence of a global ecological research project to avail itself of the best and most sophisticated specialized skill that will serve to collect data at each of the six levels. However, the choice of the parameters and of the features that bear the greatest relevance to interlevel cycling is a matter for the whole team to decide. An oscillation that goes from geological to socio-economic data-gathering is a perilous one. In such a study, centred upon the new Montreal Airport, where the Federal Government has expropriated not only the necessary operational zone but also a vast buffer area, there arises a great opportunity to test new concepts of analysis, new techniques of integration leading to a new ecological synthesis. This experiment has other dimensions, however, since it allows an actual participation of scientists in a socio-economic project of great magnitude.

Being involved as I am in this study I could be tempted to describe it in great detail. But I have done this elsewhere, and will be content to outline its procedures only inasmuch as it exemplifies the scope of ecological research on the widened scene that it occupies today. It may be useful, for instance, to enumerate some of the skills which are indispensable to such an undertaking, and, at the same time, to define some of the attitudes that are necessary to the common task.

Specialists, not generalists, must occupy the commanding position at each level: for instance, a geologist, a plant ecologist, an animal ecologist, a human geographer, an engineer or an architect, an anthropologist or a social psychologist or an economist. If possible each one of these must be assisted by other specialists. But the key men are most likely to be conversant at least in neighbouring disciplines, to the extent that one or more of them can also qualify as generalists.

The strategy in which they are mutually engaged enables them to fully test the *potential* of the land. After inventories have been made of existing situations, the plant ecologist, the agriculturist, the recreation specialist are called upon to measure capability.

And what is capability? How does one know what a landscape's *potential* is? And to what extent is this quality tied to the observer's wish? The Canada Land Inventory has applied its own system and it is inescapably based upon ecological principles. For instance, it is the qualities of climate and soil (our number I or minerotrophic level) that determine what kind of natural vegetation can be expected. Thus the fluvio-glacial sandy terraces are likely to harbour pine forest, whereas the boulder-studded moraines seem more hospitable to sugar maple forest, and the imperfectly drained clay plains will support elm, silver maple, and black ash. In turn, these sites are unevenly appropriate for different kinds of farming: the sandy terrace is excellent for horticulture, the clay flats, once drained, are very good for dairy-farming, but the moraines, once stripped of their tree cover, can only serve as very inferior pasture and are best exploited for wood and maple sugar. Likewise the potentials for wildlife, recreation, industry, and residence are viewed by the ecologist in terms of their underlying resources.

Mapping the landscape to show the location and extent of high and low potentials for these various land-uses is accompanied by warning signals that show the inhibitions: excessively rough or soft terrain, steep slope, low fertility, bad drainage, etc., all variously counter-indicate high quality agriculture or sound construction. Beyond such physical indices, however, lie the cultural requirements of recreation, hunting and fishing, urban development, and agriculture. Even if the potential is physically high, the lack of horticultural tradition in the local population may prohibit the development of strawberry gardens on the sand terraces and apple orchards on the moraines; aversion for the flesh of sheep will leave the old pastures untended. The ecologist and the planner are bound to integrate their findings by basing their summary first on the objectively definable parameters such as water availability, plant and animal productivity, adaptable investments, and checking them finally against psychological perceptions and socio-economic values.

A great variety of inscape/landscape patterns is revealed by any such analysis. The form, shape, colour, and size of widespread plants like pansies and roses, of cattle, horses, dogs, and cats all reflect a genetic-climatic patterning that is channelled by cultural forces. The bulldog, the dachshund, the husky, the borzoi, the chihuahua respond as much to visual and mythical requirements as they do to climatic tolerance and ecological functions. Notoriously the art of gardens reflects the predominantly cultural impact upon vegetal and mineral arrangement. The Japanese respect for stone as well as plant and bird has designed endless variations on seemingly very austere themes. Rigorous French geometry has cast many kinds of plants into impressive formal tapestries, whereas English devotion has preserved the natural shapes of trees and shrubs, and given rhythmic space to herbaceous borders. Arab cloisters emphasize water and stone more than plants.

Cultural heritages of this kind can be very persistent. Gilberto Freyre has amply documented the dedication of Brazilians to Mediterranean architecture, gardening, and habits in a tropical environment. The city of Christchurch, New Zealand, on the edge of a great straw-coloured tussockland and in view of evergreen beech forests, is not only traversed by a river called the Avon but is planted with English elms, birches, rose-bushes, and lilacs, and endless green lawns.

Such signs are too obvious to be missed. But how should we read our contemporary landscapes and the inscape templates that serve to model them in an increasingly uniform way? How does Canadian man, for instance, perceive wild and rural land, industrial and suburban development, and the urban environment itself? We can ask ourselves what goes on at the sixth level, in other words what controls our landscapes. We can then question whether certain malfunctions (such as the population explosion, urban sprawl, and pollution) will induce a permanent disorder. And finally, we may well ask whether there is any hope of achieving a new equilibrium by a shift in values, and therefore a new series of prototypic inscapes.

Much has been made of the pioneer spirit of Canadians, of their frontier mentality, of the live heritage of the "coureur de bois" that inspires the insurance salesman in his autumnal stalkings and his wife-escaping fishing trips. Certainly, to this day, the Canadian city-dweller does hanker after sparkling trout streams and reed marshes with settling ducks and open brushland with leaping deer. He wants his wild sporting grounds preserved and, if possible, to own a cabin or a cottage near a lake or a river. Such enjoyments are available, in fact, to a fairly large proportion of the population, even at low income levels. Enjoyment of nature is a strong factor, and social status another. In such a remote and seemingly wild territory as the Magdalen Islands, out in the Gulf of St. Lawrence, where virtually all houses are within sight of the water, many families have their seaside "chalet". In one extreme case, such a dilatory domicile (as the New York Social Register calls it) was built next door to the owner's permanent home!

In the authenticity and strength of their need for contact with nature, I doubt that the Canadians can nearly compare wih the New Zealanders who are, to a man, participants in hiking, who keep up their woodland tracks ever so carefully and practise every kind of outdoor exercise. Their knowledge of plants and wildlife, their sense of recognition in the rhythm of the seasons, and the events triggered off by biological change, are very acute and ready to emerge at the level of consciousness.

There is a great deal of wilderness left in Canada and some of it is both well protected and accessible, but much of it is badly spoiled without necessarily being profitably exploited. Clearcut hillsides, sloppily dammed rivers, gaping gravel pits with pools of stagnant water, are neither wild, nor rural, nor industrial, nor urban, but just plain *waste land*. This "agony in stony places" does not elicit strong protest and efficient reparation or correction, although there are signs that consciousness is awaking.

Possibly the rural landscape bears more reality for the Canadian than the wilderness. Although the rural–urban balance of population has been tipped in favour of the cities a generation ago, this statistical fact does not bear the implication of an overwhelming urban mentality. It is quite possible that most Canadians are city-dwelling rurals. Their way of life, if compared with that of Londoners, Parisians, Berliners, retains a great deal of the parochial behaviour one associates with the smaller agglomerations. The myth of the "goodness" of country life is in great part responsible for such persistent institutions as the electoral divisions that so often allow one farmer to cancel out five or six urban votes. This need not imply anything like a bucolic view of rural life. Our literature presents strong contrasts in this respect, for instance Lucy Maud Montgomery's sweet *Anne of Green Gables* at one end and Phillippe Panneton's sad *Thirty Acres* at the other.

The flight from rural boredom and drudgery to urban–industrial comfort and frustration has engaged large masses of people, in Canada as elsewhere. This exodus has been examined more closely by economists, historians, and sociologists than by ecologists. It has been said that the progress of technology was the major determinant, since it reduced the need for a large labour force on the farm and correspondingly absorbed a larger one in the city, that is until further technology reached the level of widespread automation.

In our present context, the questions that we ask ourselves, having defined the wild, rural, industrial, and urban ecosystems in terms of the way in which their trophic levels are occupied and the extent to which they depend upon other ecosystems, are the following:

1. What energy transfers are involved in the succession, in a given landscape, from wild to rural to industrial or urban?

2. What underlying intentions (or system of values) are the basic motive of our management practices?

3. What is the present perception of existing landscapes that sustains the forms of cycling that we now witness?

4. What changes can we foresee in human motivation and interaction that promise hope for a better environment?

Maybe enough has already been said concerning the under-pinnings of resource strategy to provide a clear frame for the first question. Briefly, a landscape in the Montreal Plain, when it passed from the wild to the rural stage, was dammed and drained, stripped of much of its vegetal cover, rid of most of its wildlife, ploughed, sown, planted, harvested, and grazed. Little was left to the sway of age-old ecosystems; short-term primary and secondary production units were set up under constant control. When the farm, in its turn, was claimed by industry or urbanization, its soil was sealed up, its exotic plants and animals replaced by factories, stores, and houses where processing and services were effected by a human population of unprecedented density, dependent upon outside resources but also powerful in obtaining high-value investments in exchange for its products.

In answer to the second question, if we try to seize the nature of the drive that accompanied the transformation of woodland and marsh to field and orchard and on to factory and row-houses, we discern several intentions. One is certainly the increasing physiological safety of humans as far as meteorological and animal hazards are concerned. Another is the growing capacity to acquire objects that the local ecosystem does not or cannot produce. Yet another is the freedom from metabolic and survival tasks and the accomplishment of others that are presumed materially, economically, socially, and maybe spiritually more rewarding. How early the idea of what we call progress may have set in is hard to say. Barbara Ward (1961) has described it in its industrial and post-industrial form. Ignazio Silone, who belongs to the generation in which marxists and capitalists agreed on the development of greater wealth, has recently made some disenchanted statements on this topic. My concern, for the moment, is more ecological than ethical, however, and I can only recognize as a driving force these various motivations that promoted landscape succession.

Maybe, in answer to the third question, we can find access

to the ethical plane, and decide that landscape management is not amenable to a world policy in any other terms.

What then, and this is the third question, has brought about the present disorder of our world, the present spoilage of our landscapes, and what maintains them in that state? The principal forces, as I see them, are: population increase, industrial growth, and poor management practices.

Some authors (e.g., Dumont and Rosier 1966) have expressed complete despair as far as some countries, such as India and Haiti, are concerned. There, the eventual capacity of the land to feed the already excessive *population* is beyond the ken of any conceivable technological progress, and even better sharing of existing resources will not prevent famine in the near future. One may have reservations about such pessimistic forecasts, but cannot deny that the present population trend has to be curbed in the developed as well as in the developing countries. In spite of my diffidence towards militant attitudes, I do not hesitate to side with Zero Population Growth. I simply cannot believe that new technological advances are capable of keeping up with continued increase in the numbers of the human species, even if some built-in biological controls stop us short of what Karl Sax has called *Standing Room Only* (1960). I need hardly recall the bitter denunciations by Ehrlich (1971), Commoner (1972), and so many others to underline the extreme urgency of this great problem and its attendant moral anguish. We could well pause to reflect on the relative optimism of physicists and engineers and the pessimism of biologists on the resource-population race. Is this to be read as a historical development, possibly as a cultural accident? Or is it rooted in the very nature of the physical vs. the biological disciplines and their possibly divergent optical process?

Industrial growth is the second cause of resource depletion, of impending shortages of fuel, and of energy. It is also the cause of pollution and spoilage of the environment by widespread encumbrance of irreducible substances, and of urban sprawl. Finally, it has erected itself into a mythical position as the very promise of earthly happiness. Serious proposals for the slowing down of industrial productivity may not have been triggered off by the hippies who turned their backs upon the consumers' society, but I am confident that the rejection of many morally untested and untenable myths of the industrial age has given heart to well-qualified economists and social scientists in their denial of accepted standards. Only yesterday nothing seemed more reliable than the Gross National Product as a yardstick of national prosperity. The great economist Kenneth Boulding's irreverent "Fun and Games with the Gross National Product" (1970) drew attention to "the role of misleading indicators in social policy". Dennis Meadows and his collaborators went several steps further in their *Limits to Growth* (1972), and offered proof not so much that we *must* not continue to grow as we now do, but that we *can* not! The Stockholm Conference on the Human Environment and its "Declaration", as well as the "Blueprint for Survival" of the British *Ecologist*, make these issues more amenable to decision and their challenge cannot possibly be turned aside. The controversy is now too open to be denied, although the Stockholm document bears witness to some disquieting compromises (Allen 1972).

I shall not analyze this trend, even less judge how much we can take and how much we can leave, and if we should go all the way or only part of the way with the advocates of reduced production rate. Within the framework of my present topic, I find it heartening that economists and others who are by vocation quantitatively minded are seeking ways to ponderate certain values (beliefs, perceptions, behaviours which have been considered in the present essay for instance) that have generally escaped quantitative estimate but are so evidently capable of stimulating or counteracting forces (fuel supply, population growth, consumption of food) which are amenable to quantification.

On a parallel with population growth and industrial growth

is a third force which I am calling *management practice*. This literally covers a multitude of sins, for I am referring to that power that only some nations have of controlling resources and meting out their benefits. On a world basis "the rich and the poor nations" have achieved disastrously different levels of living standards. The present distribution is the result not so much of the actual location of the resources themselves as it is the outcome of migrations, wars (hot and cold), and of national and international conspiracies. Whatever the range of our conspectus (Appalachia, Arctica, or Suburbia), the scale from starvation to conspicuous spending and wanton waste can be witnessed. Political, economic, cultural, and religious levers control the spread of both resources and accumulated wealth.

This normally introduces the fourth and last question which concerns the future. If it is true, as I have argued, that we have managed our wild, rural, industrial, and urban landscapes by making them yield the goods which we most value and have suffered consequences which we now find that we can no longer safely accept, what can we do? I shall not present a fully drawn blueprint for a joyous austerity, certainly not as a prescription for the definitive happiness of mankind on earth. But I do believe I can point to areas of reform in the planning process that require a shift in some basic attitudes and values.

A sharing of resources which allows their relatively short-term renewal and which provides at least minimum gratification of man's needs is compatible neither with the present rate at which population is growing and resources are being tapped, nor with the present allocation of power to use and to store them. I cannot and do not wish to avoid either ethical or political issues and their underlying carrier of economic disparity. I hope I can continue to look upon psycho-political forces as ecological levers. Although I claim to have no commitment to any political party nor to any specific ideology, I recognize that my sense of justice is of Christian inspiration.

For the rest I shall not further qualify my statements, trusting that they will be honestly considered as a partial contribution for what consistency and perceptivity they may possess.

What I now wish to attempt, in concluding this essay, is an ultimate scaling of the ecosystem levels, as defined, with the object of exploring the forces in our landscapes and in our minds that can possibly permit this joyous austerity in which alone I foresee worldly salvation. This ladder is not without its angels to be fought as best I can at each rung so as to reach the next and eventually the highest point.

In an earlier essay (1971), I had made an attempt to explore the ecological basis of human rights, and even to draw a tentative chart that recognized the conflicts between rights and duties, opportunities and actual use by examining first the *needs* of the individual (physiological, psychological, social, economic, and ethical), the *rights* of the encompassing societies, and finally the basic *responsibility* of individual and society to the human *species* as a whole. I shall not cover this ground as thoroughly again, but will beam my attention on some of the points where the claims of the individual, the response of society, and the international disparity are so much at odds that our very survival, or at least our peaceful survival, is at stake.

Considering first the mineral resources, the price which we have to pay for light, air, and water can hardly be too high. The increase of sunlight hours and the return of fish in the Thames River in London prove that long-standing pollution problems can be solved by restoring mineral cycling to its normal channels. Getting used to pollution is not the answer, although it is only too current. A colleague of mine, born in the ugliest town in Canada—Sudbury, Ontario—says that he likes the fume-laden air of that moonlike landscape. And who was that "grande dame" who visited Versailles after the revolutionary clean-up and rather missed the stench that the French Court so readily blended with its rare perfumes? No, conditioned reflexes will not do! Tolerance of conditions harmful to physical and mental health cannot be accepted as a necessary evil. Poisoning of air, soil, and water is a malfunction in many related ecosystems, and corrections are urgently required.

At the level of man's needs in plant and animal nutrients, I am sorry I cannot subscribe to the naturists' insistence on so-called natural foods, although they are perfectly right in their strictures against pesticide poison accumulations in vegetable and animal tissues. The consumption by many urban and even rural people of processed food and ready-made dishes also leads to a great loss of awareness and of enjoyment. One need hardly be a gourmet to realize that good food satisfies all the senses and indeed the intellect, and is a carrier of cultural communication. It is no accident that the best cooking the world has known so far, the Chinese and the French, is supported by a tradition of thrift and curiosity. A knowledge of the shape, colour, and growth-form of all varieties of vegetables and of the exact anatomy of animals, and the wide repertory of both raw materials and recipes in these two cultures point to environmental consciousness even more than they indicate a high level of sophistication. I shall make no invidious comparisons with other civilizations where eating is primarily a fueling operation. It is, however, worth looking back (if that is the word) to the ritual slaying of caribou by Quebec Indians who elaborately invoke the consent of the live animal before sacrificing him. It is just such ways of decompartmentalizing our behaviour that are needed in post-industrial society. Consumption of food is not often enough perceived by us as a celebration unless the occasion is conventionally identified (Sundays, birthdays, holy days).

There are plenty of frustrations in modern eating and drinking habits, but even more in the various investments that give shape, form, and function to our collective way of life. How many of us live in houses that we really like? Not even all of those who can afford them. How many have enough space, silence, safety? How many have achieved a satisfactory ratio of work and play, of movement and retreat, of privacy and social contact? I can only sample a few of the issues these questions of human investment raise.

In this respect, even more than on the lower levels of mineral and biological supply, the lag is great between reality, as scientifically observed in the ecosystem, and perception by individuals and whole societies. I should like to concentrate on *income, housing,* and *transport,* as I think they provide the best clues to this discrepancy and pose the most crucial questions concerning the mismanagement of the environment.

Purchasing power that enables the individual or the family to obtain goods and services they cannot provide for themselves is the reward for their own contribution to the pool of exchangeable products. The latter undergo fluctuating values mostly unrelated to the energy expenditure in production. The way in which income levels are attained is controlled by accidental historical structures that grossly abuse the long-term potential of world ecosystems. Disparities in the Himalayas, the Appalachians, and the Andes have settled into fixed patterns through the play of violence and privilege. The vast conspiracy of advertising, using the great arsenal of hidden persuasion and triggering chain-reactions of conditioned reflexes, has successfully topped natural needs with a number of secondary and often burdensome cravings that sorely tax the working and earning capacity of the individual. This overloading is the dark side of the consumers' society. One need subscribe to no particular political ideology to recognize the disorder engendered by the present social compulsion to buy and to show, and to overconsume, and eventually to waste. One need hardly ask if this is moral, if it is responsible. One can pose the more pragmatic question: "Can it continue?" If the consumers' society is based on growth, are there no limits to growth? Ecologists all say that there are, even if they cannot agree in identifying the parameters that will light up the end-points, and quantify the tempo.

Redistribution of income within a society (through taxation, welfare benefits, etc.) is not being handled very well any-where. Compensation between societies is regressing instead of progressing. Proof is constantly added to proof that the rich are getting richer and the poor poorer. But it seems that even a more equitable spread would not solve our long-range problems and that a slow-down of production and consumption may soon be considered as essential as the reduction of the rate of human population growth. This points to sacrifice, to austerity of a kind.

Housing and, in a broader sense, space occupied and controlled by the individual and the family raise so many economic, esthetic, social, and psychological issues that I shall not take time to enumerate them. As a participant in two intensive exercises in this field (with the Federal Task Force on Housing and Urban Development (1968-69) and with the Science Council of Canada) I have been given a glimpse of what Canadians have and of what they want and occasionally of what they think they want. As co-author of their published reports, I have been given the opportunity to define the problems and to recommend a partial course of action. As often as not social attitudes present as large a stumbling block as do economic forces and jurisdictional confusion.

For one thing, the psychology of ownership could very well be at odds with the economic and social realities of the present. Almost everyone would like a privately owned, somewhat isolated one-family house, with a garden. The implication sometimes arises that this is "natural", that it is "best", and occasionally that it is even a right. Attempts are being made at present to analyze this expressed desire, to submit its components (architectural, spatial, social, economic, cultural) to close scrutiny, in order to decide whether the amenities of familial harmony which are aimed at really depend upon the proposed components. For instance, what is the weight of private ownership? It is generally perceived as conferring total freedom to build, alter, decorate, furnish, and use. This concept may very well have been objectively correct a few generations ago for a privileged group of people. Its actual base is highly questionable in a civilization of high taxes on property, succession, and income, long-term mortgages, costly and questionably competent repairs, changing zoning laws, and fluctuating assessments and real-estate values. There is little sense in arguing these points, however, with many people (proprietors and non-proprietors) whose cultural fixation is of the nature of moral certainty and sociopolitical commitment. For instance, some well-to-do farmers, on good land that is expropriated for an airport development but not affected by the operation itself, and who are well compensated, and offered the continued and uninhibited use of their land at a low rental, move away in sheer frustration. The withdrawal of the legal deed has the strength of a flood or a hurricane in alienating them from their life's work in this particular environment. Maybe usufruct is the real issue but they regard ownership as more vital.

Other areas that lend themselves to this kind of ecological approach are work-and-play, education and information, participation in collective decisions, etc. But I shall consider only one more: transportation, with special reference to the automobile.

At cocktail parties and business meetings, gatherings of discothecous young people, and in many other ritual or casual groupings, the conversation may dwell for hours on motorized transport: plane, automobile, motorcycle, skidoo, or motorboat. The glare on the covers of drugstore magazines comes as strongly from the goggle eyes of headlights as from the bobbing breasts of beauties. The perception of motorized transport forces its way into every single phase of everyday life, even in the rural environment. The inevitability of private transport by automobile and of private ownership (maybe on credit) has not really been questioned. The planning and architecture of our suburbs are entirely geared to this idea. Old and once-beautiful cities like Paris and London are glutted with cars in spaces never meant for parking. The deteriorating ratio of public to private transportation has even made inroads in some of these European cities, traditionally given to the noble art of walking and to the congenial customs of public vehicles. The redirection of traffic through one-way streets, severe and expensive parking regulations, and the development of malls do not keep up with the rising tide of automobiles which all too often carry only one passenger. New York now lies beyond the limit of tolerance; Toronto and Montreal are approaching it rapidly. The growth of the number of automobiles must be stopped. Prohibition of private vehicles in town centres must be instituted soon. Crowding, pollution, waste, frustration, noise, danger: the automobile has not replaced the horse as the most noble conquest of man.

Austerity is coming. Scarcity, social and economic stress are closing down upon us. It is later than we think.

Is there any hope that this unwise, spendthrift civilization has seen the handwriting on the wall, or is there too much grime upon it for the letters to be read? If we are not utterly corrupted, what sources can we tap so as to give this necessary austerity a joyous character? How can we, in Buckminster Fuller's terms, "do more with less?"

I could possibly be very eloquent if I said that I put my faith in the young; that their rejection (far from unanimous, for many of them are scroungers, as in the past) of the consumers' world and of war promises a return to the simple pleasures of frugal eating, primitive shelter, and unlimited love-making. But I have always felt that such statements are a courting of tomorrow's fashions and a denial of healthy social structure. The richness and the variety of experience of the several generations that share almost any environment at any time are indispensable to the proper cycling of information. The loss involved in the stamping out or the inhibition of an age-class or an occupational group is irreparable and leaves too much wealth untapped.

How can we ultimately test the moral fibre of a society

that will face up to the inevitable loss of resource potential, to the eventually unbearable burden of population growth and industrial productivity, and will decide to curb itself, to revise its common goals and knowingly design its future? The joyous austerity which I dare to think of can only be based on consent. B. F. Skinner (1972) has just pointed out which portions of our cake we can still hope to consume and what verbal illusions of "freedom and dignity" it is now unsafe to cherish. Whereas such a resolution would have to be widespread, is it too much to hope that Canada will launch upon a course of leadership in this respect? The dangers of unilateralism are well known to all of those who opt for the common good. But noble risks are the only ones worth taking.

I realize how open-ended many of my propositions are. Those who have more talent for questions than for answers badly need to work with their opposite numbers who are more immediately anxious for solutions. It is in the hope of such exchanges that the present attempt at a general ecological design has been offered.

Aesop (ed. by James E. Wetherell, 1926). *Aesop in Verse.* Macmillan, Toronto, xvi + 199 pp.

Allen, Robert. 1972. "Can Stockholm Survive New York?" *The Ecologist,* 2(10): 4–9.

Arber, Agnes. 1954. *The Mind and the Eye.* Cambridge Univ. Press, Cambridge, Mass., xi + 146 pp.

Audubon, John James. 1961. *The Birds of America.* (In part reprinted from 1827–30 volume.) Macmillan Co., New York, xxvi + 435 pp.

Bates, Marston. 1964. *Man in Nature.* (Second edition.) Prentice-Hall, Englewood Cliffs, New Jersey, x + 116 pp.

Baulig, Henri. 1956. *Vocabulaire franco-anglo-allemand de géomorphologie.* Publ. Fac. Lettres, Univ. Strasbourg, Fasc. 130, xiv + 230 pp.

Bernard, Claude. 1865(1961). *An Introduction to the Study of Experimental Medicine.* (Transl. by Henry Copley Greene.) Collier Books, New York, 255 pp.

Boulding, Kenneth. 1964. "The Meaning of the Twentieth Century". *World Perspectives,* Harper & Row, New York, Vol. 24, xvi + 199 pp.

———. 1966. "Economics and Ecology". In: *Future Environments of North America,* ed. by F. Fraser Darling and John P. Milton, Natural History Press, Garden City, New York, pp. 225–234.

Boulding, Kenneth. 1970. "Fun and Games with the Gross National Product—the Role of Misleading Indicators in Social Policy". In: *The Environmental Crisis*, ed. by H. W. Helfrich, Jr., Yale Univ. Press, New Haven, pp. 157-170.

Brion, Marcel. 1955. *Les animaux, un grand thème de l'art.* Horizons de France, Paris, 44 pp. + 100 planches en noir et blanc + 16 planches en couleurs.

Brunet, Michel. 1958. *La présence anglaise et les Canadiens.* Beauchemin, Montréal, 293 pp.

Brunhes, Jean. 1910(1920). *Human Geography: An Attempt at a Positive Classification.* (Transl. by I. C. LeCompte.) Rand-McNally, Chicago, xvi + 648 pp.

——. 1925. "Human Geography". In: *The History and Prospects of the Social Sciences*, ed. by H. E. Barnes, Knopf, New York, pp. 55–105.

Burgess, T. W. 1920. *The Adventures of Jimmy Skunk.* Little, Brown & Co., New York, 118 pp.

Cain, Stanley A. 1950. "Natural Resources and Population Pressures". *Bios* 21(4):247–259.

——. 1951. "The Interdependence of Man and Natural Resources". *Proc. 2nd Regional Conf. on Conserv. of Nat. Resources*, pp. 20–29.

——. 1953. *The First Three Years.* Univ. Mich. School of Nat. Res., Dept. Conserv., 69 pp.

——. 1963. *Syllabus: Natural Resource Ecology.* Univ. Michigan School of Natural Resources, n.p.

——, et al. 1969. *Man and Nature in the City.* Symp. Bur. Sport Fish. & Wildlife, U.S. Dept. Interior, Washington, Oct. 1968, xii + 92 pp.

Canada. 1965(1970). *The Canada Land Inventory. Objectives, Scope and Organization.* Dept. Regional Econ. Expansion, Canada Land Inventory Report No. 1, iv + 61 pp.

——. 1965 (1969). *The Canada Land Inventory. Soil Capability Classification for Agriculture.* Dept. Regional

Econ. Expansion, Canada Land Inventory Report No. 2, 16 pp.

———. 1966. *The Canada Land Inventory. The Climates of Canada for Agriculture.* Dept. For. & Rural Devel., Canada Land Inventory Report No. 3, vi + 24 pp. + maps.

———. 1969. *The Canada Land Inventory. Land Capability Classification for Outdoor Recreation.* Dept. Regional Econ. Expansion, Canada Land Inventory Report No. 6, ii + 114 pp.

———. 1970. *The Canada Land Inventory. Land Capability Classification for Wildlife.* Dept. Regional Econ. Expansion, Canada Land Inventory Report No. 7, v + 30 pp.

Candolle, Alphonse de. 1855. *Géographie botanique raisonnée.* Impr. de L. Martinet, Paris, 2 vols.: xxxii + 1365 pp.

Carson, Rachel. 1951. *The Sea Around Us.* Oxford Univ. Press, New York, vii + 230 pp.

Carver, Humphrey. 1962. *Cities in the Suburbs.* Univ. of Toronto Press, Toronto, viii + 120 pp.

Castro, Ferreira de. 1930(1954). *A selva.* Guimarães & Cª., Lisboa, 322 pp.

Clark, Andrew H. 1949. *Invasion of New Zealand by People, Plants and Animals. The South Island.* Rutgers Univ. Press, New Brunswick, New Jersey.

Clements, Frederic E. 1949. *Dynamics of Vegetation.* H. W. Wilson Co., New York, 23 + 296 pp.

Commoner, Barry. 1971. *The Closing Circle. Nature, Man and Technology.* Alfred A. Knopf, New York, x + 326 pp.

Cowles, Henry Chandler. 1899. "The Ecological Relations of the Vegetation on the Sand Dunes of Lake Michigan". *Bot. Gaz.*, 27(2–5):95–117, 167–202, 281–308, 361–391.

Dansereau, Pierre. 1945. "Les conditions de l'acériculture". *Agriculture*, 2(1):45–47; (2):140–152; (3):259–267; (4):235–291; also *Bull. Serv. Biogéogr.*, No. 1, 52 pp.

———. 1966. "Ecological Impact and Human Ecology". In: *Future Environments of North America*, ed. by F. Fraser

Darling and John P. Milton, Natural History Press, Garden City, New York, pp. 425–462.

————. 1969. "The Hope of Human Ecology". *Bull. Canad. Comm. for UNESCO*, 12(1–2) Suppl., 14 pp.

————. 1970. "Megalopolis: Resources and Prospect". In: *Challenge for Survival: Land, Air, and Water for Man in Megalopolis*, ed. by Pierre Dansereau, Columbia Univ. Press, New York, pp. 1–33.

————. 1970. "Ecology and the Escalation of Human Impact". *Int. Soc. Sci. Jour.*, 22(4):628–647.

————. 1970. "Reflections on a Task: Housing and Urban Development in Canada, 1968". *Sarracenia* No. 13, 42 pp.

————. 1971. "Dimensions of Environmental Quality". *Sarracenia* No. 14, 109 pp.

————. 1971. "EZAIM—an Interdisciplinary Adventure". *Natl. Res. Counc. Canada Newsletter*, 3(3):4 pp.

———— (ed.). 1971. "Cities for Tomorrow. Some Applications of Science and Technology to Urban Development". *Science Council of Canada Report* No. 14, Preface, pp. 4–7; also *PLAN*, 11(3):247–250. 1972.

————. 1972. "Biogéographie dynamique du Québec". In: *Etudes sur la Géographie du Canada: Québec*, éd. par Fernand Grenier, Univ. Toronto Press, Toronto, pp. 74–110.

Darling, F. Fraser. 1964. "Conservation and Ecological Theory". *Jour. Ecol.*, 52(Suppl.):39–45.

————, and John P. Milton (eds.). 1966. *Future Environments of North America*. Nat. Hist. Press, Garden City, New York, xv + 767 pp.

Darlington, C. D. 1970. *The Evolution of Man and Society*. Allen & Unwin, London.

Darwin, Charles. 1859(1898). *The Origin of Species by Means of Natural Selection*. D. Appleton & Co., New York, Vol. I:xxvi + 365 pp.; Vol. II:vii + 338 pp.

Dickens, Charles. 1837–38. *Oliver Twist*. Illustrated Sterling Edition, Dana Estes & Co., Boston, vii + 418 pp.

Dobzhansky, Theodosius. 1950. "Evolution in the Tropics".
Amer. Scientist, 38(2):209–221.

———. 1967. *The Biology of Ultimate Concern.* New American Library, New York, xviii + 152 pp.

Doxiadis, Constantinos. 1968. *Ekistics. An Introduction to the Science of Human Settlements.* Hutchinson, London, xxix + 527 pp.

Dumont, René, et Bernard Rosier. 1966. *Nous allons à la famine.* Editions du Seuil, Paris, 280 pp.

Duvigneaud, P. 1963. "Ecosystèmes et biosphère". Volume 2 de *L'Ecologie, science moderne de synthèse.* Min. Educ. Natl. et Cult., Bruxelles, *Documentation 23,* 130 pp.

Ehrlich, Paul. 1968. *The Population Bomb.* Ballantine Books, New York, 223 pp.

———, and Anne H. Ehrlich. 1970. *Population, Resources, Environment. Issues in Human Ecology.* W. H. Freeman & Co., San Francisco, 383 pp.

Eliot, T. S. 1922. "The Waste Land". In: *Collected Poems 1909–1935,* T. S. Eliot, 1936, Faber & Faber, London, pp. 61–84.

Emerson, Ralph Waldo. 1836. "Nature". In: *Collected Works of Ralph Waldo Emerson,* Greystone Press, New York, pp. 337–363.

Falardeau, Jean-Charles. 1967. "Notre société et son roman". Editions HMH, Montréal, *Science de l'homme et humanisme,* No. 1, 234 pp.

Farvar, M. Taghi, and John P. Milton (eds.). 1972. *The Careless Technology: Ecology and International Development.* Doubleday & Co., Inc., Garden City, New York, 1060 pp.

Freyre, Gilberto. 1941. *Região e tradição.* Livraria José Olympio, Rio de Janeiro, 264 pp.

———. 1946(1956). *The Masters and the Slaves.* (Transl. by Samuel Putnam.) Alfred A. Knopf, New York, 2nd edit., lxxi + 537 + xliv pp.

Fuller, Buckminster. 1963. *Ideas and Integrities; A Spontaneous Autobiographical Disclosure.* Collier-Macmillan Canada Ltd., Toronto, 318 pp.

Galbraith, John Kenneth. 1967. *The New Industrial State.* Houghton-Mifflin, Boston, xiv + 427 pp.

Glacken, Clarence J. 1966. "Reflections on the Man-Nature Theme as a Subject for Study". In: *Future Environments of North America*, ed. by F. Fraser Darling and John P. Milton, Nat. Hist. Press, Garden City, New York, pp. 355–371.

Goldsmith, E., *et al.* 1972. "A Blueprint for Survival". *The Ecologist*, 2(1):1–43.

Gothein, Marie Luise (ed. by Walter P. Wright). 1928. *A History of Garden Art.* Volumes I and II. J. M. Dent & Sons Ltd., London and Toronto, E. P. Dutton & Co. Ltd., New York, I:xxiv + 459 pp.; II:xv + 486 pp.

Grandtner, Miroslav M. 1966. *La végétation forestière du Québec méridional.* Presses Univ. Laval, Québec, xxv + 216 pp.

Hahn, E. 1896. *Die Haustiere und ihre Beziehunger zur Wirtschaft des Menschen.* Duncker & Humblot, Leipzig.

Hardy, Thomas. 1878(1966). *The Return of the Native.* Harper & Row, New York, xxviii + 420 pp.

———. 1891(1957). *Tess of the d'Urbervilles.* St. Martin's Press, New York, xi + 446 pp.

Hare, F. Kenneth. 1961. *The Restless Atmosphere.* Hutchinson Univ. Library, London, i–viii + 9–192 pp.

———, and J. C. Ritchie. 1972. "The Boreal Bioclimates". *Geogr. Rev.*, 62(3):333–365.

Hautecoeur, Louis. 1959. *Les jardins des dieux et des hommes.* Hachette, Paris, 230 pp.

Hellyer, Paul T., *et al.* 1969. *Report of the Task Force on Housing and Urban Development.* Queen's Printer, Ottawa, 85 pp.

Hochreutiner, B. P. G. 1911. *La philosophie d'un naturaliste.* Félix Alcan, Paris.

Hopkins, Gerard Manley. 1953. *A Hopkins Reader* (ed. by John Pick). Oxford Univ. Press, New York, 317 pp.

Humboldt, Alexander von. 1859. *Cosmos: A Sketch of a Physical Description of the Universe.* Volumes I to V. Harper Bros., New York.

Hutchinson, George Evelyn. 1950. "Survey of Contemporary Knowledge of Biogeochemistry". 3. "The Biogeochemistry of Vertebrate Excretion". *Bull. Amer. Mus. Nat. Hist.*, 96:xviii + 554 pp.

——. 1965. *The Ecological Theater and the Evolutionary Play.* Yale Univ. Press, New Haven, xiii + 139 pp.

Huxley, T. H. 1863(1969). *Evidence as to Man's Place in Nature.* Repr.: Ann Arbor Paperbacks. Univ. Mich. Press, Ann Arbor, AA–24, 184 pp.

International Council of Scientific Unions. 1972. *International Biological Programme, 1972 Review.* Spec. Comm. for Int. Biol. Progr. (SCIBP), London, 65 pp.

Isaac, Erich. 1962. "On the Domestication of Cattle". *Science*, 137(3525):195–204.

Jeffrey, W. W., *et al.* (eds.). 1970. *Towards Integrated Resource Management. Report of the Sub-Committee on Multiple Use, National Committee on Forest Land. Principaux commentaires et recommendations.* Dept. Regional Econ. Expansion, Ottawa, xxiii + 47 pp.

Kelly, Walt. 1953. *The Pogo Papers.* Simon & Schuster, New York, 192 pp.

Kerner von Marilaun, Anton. 1863. *Das Pflanzenleben der Donauländer.* Transl. by Henry S. Conard, 1951, as *The Background of Plant Ecology.* Iowa State Coll. Press, Ames, x + 238 pp.

Kessel, Joseph. 1958(1959). *The Lion.* (Transl. by Peter Green.) Hart-Davis, London, 208 pp.

King, Lawrence J. 1966. *Weeds of the World: Biology and Control.* Interscience Press, New York.

Kipling, Rudyard. 1948. *Jungle Books.* Doubleday & Co., New York, 1:253 pp.; 2: 201 pp.

La Fontaine, Jean de. 1664–74(1954). *The Fables of La Fontaine* (transl. by Marianne Moore). Viking Press, New York, x + 342 pp.

Lamartine, Alphonse de. 1820. "Méditations". In: *Oeuvres poétiques complètes*, Bibliothèque de la Pléiade, Gallimard, Paris, No. 165, xxxviii + 2030 pp.

Leopold, Aldo. 1949(1969). *A Sand County Almanac and Sketches Here and There.* Oxford Univ. Press, Oxford, London, New York, reprint edition, xiii + 226 pp.

Leroi-Gourhan, André. 1965. *Préhistoire de l'art occidental.* Editions d'Art Lucien Mazenod, Paris, 482 pp.

Lévi-Strauss, Claude. 1962(1966). *The Savage Mind.* Univ. Chicago Press, Chicago, xii + 290 pp.

Lindeman, Raymond L. 1942. "The Trophic-dynamic Aspect of Ecology". *Ecology*, 23(4):399–418.

Lowenthal, David. 1964. "Is Wilderness 'Paradise Enow'? Image of Nature in America". *Columbia Univ. Forum*, 7(2):23–40.

———, and Marquita Riel. 1972. *Publications in Environmental Perception.* Nos. 1–8. American Geographical Society, New York.

Lowie, Robert H. 1963. *Indians of the Plains.* Amer. Mus. Sci. Books, Nat. Hist. Press, Garden City, New York, xx + 258 pp.

Lynch, Kevin. 1960. *Image of the City.* Mass. Inst. Technology Press, Cambridge, Mass., vii + 194 pp.

Malin, James C. 1956. "The Grassland of North America: Its Occupancy and the Challenge of Continuous Reappraisals". In: *Man's Role in Changing the Face of the Earth*, ed. by Wm. L. Thomas, Univ. Chicago Press, Chicago, pp. 350–366.

Malthus, T. R. 1798(1959). *Population: the First Essay.* Ann Arbor Paperbacks, Univ. Mich. Press, Ann Arbor, AA–31, vi + 139 pp.

McLuhan, Marshall. 1964. *Understanding Media.* McGraw-Hill Book Co., New York, Toronto, vii + 360 pp.

McNeil, Raymond, 1969. "La détermination du contenu lipidique et de la capacité de vol chez quelqu(s espèces d'oiseaux de rivage (Charadriidae et Scolopacidae)". *Can. Jour. Zool.*, 47(4):525–536.

———. 1970. "Les grands voyages des oiseaux./Le carburant des oiseaux migrateurs./Navigation et orientation des oiseaux migrateurs". *Québec-Science*, 8(6):25; (7):7–9; 9(2):3–6.

Meadows, Donella H., *et al.* 1972. *The Limits to Growth.* Universe Books, New York, 205 pp.

Montgomery, Lucy Maud. 1908(1968). *Anne of Green Gables.* Ryerson Press, Toronto, vi + 329 pp.

Mumford, Lewis. 1966. *The Myth of the Machine. Technics and Human Development.* Harcourt, Brace and World, New York, viii + 342 pp.

Nicholson, E. Max. 1964. *Conservation and the Next Renaissance.* Univ. Calif. School Forestry, Horace M. Albright Conservation Lectureship IV: 1–15.

———. 1968. *Handbook to the Conservation Section of the International Biological Programme.* IBP Handbook No. 5, Blackwell Sci. Publ., Oxford, ix + 84 pp.

———. 1970(1972). *The Environmental Revolution. A Guide for the New Masters of the World.* A Pelican Book, Penguin Books, England, 432 pp.

Odum, Eugene P. 1953. *Fundamentals of Ecology.* W. B. Saunders Co., Philadelphia, xii + 384 pp.

Odum, Howard T. 1971. *Environment, Power, and Society.* Wiley-Interscience, New York, ix + 331 pp.

Orwell, G. 1946. *Animal Farm.* Harcourt, Brace & Co., New York, 118 pp.

Panneton, Philippe (Ringuet). 1938(1940). *Thirty Acres* (transl. by Felix and Dorothea Walter). Macmillan, Toronto, 342 pp.

Pascal, Blaise. 1910. *Thoughts, Letters and Minor Works.* P. F. Collier & Son, New York, Harvard Classics Vol. 48, 451 pp.

Piaget, Jean. 1967(1971). *Biology and Knowledge*. (Transl. by Beatrix Walsh.) Univ. Chicago Press, Chicago, xii + 384 pp.

Pitseolak. 1972. *Pictures Out of My Life. Tape-recorded Interviews by Dorothy Eber*. Design Collaborative Books, Montreal, and Oxford Univ. Press, Toronto, n.p.

Richards, P. W. 1952. *The Tropical Rain Forest. An Ecological Study*. Cambridge Univ. Press, xviii + 450 pp.

Saarinen, Thomas F. 1969. *Perception of Environment*. Ass. Amer. Geogr., Comm. on Coll. Geogr., Resource Paper 5, 37 pp.

Saint-Exupery, A. de. 1955. *Wind, Sand and Stars*. (Transl. by Lewis Galantière.) Heinemann, Toronto, 285 pp.

Salten, Felix. 1928(1939). *Bambi*. Pocket Books Inc., New York, 209 pp.

Sauer, Carl O. 1952. *Agricultural Origins and Dispersals*. Bowman Mem. Lecture, Amer. Geogr. Soc., New York, Ser. 2, v + 110 pp.

Sax, Karl. 1955(1960). *Standing Room Only*. Beacon Press, Boston, xviii + 206 pp.

Schwanitz, Franz. 1966. *The Origin of Cultivated Plants*. Harvard Univ. Press, Cambridge, Mass., vi + 175 pp.

Sears, Paul B. 1942. "History of Conservation in Ohio". In: *History of the State of Ohio, VI, Ohio in the Twentieth Century*, Ohio State Archeol. & Hist. Soc., Columbus, pp. 219–240.

———. 1947. "Importance of Ecology in the Training of Engineers". *Science*, 106(2740): 3 pp.

———. 1954. "Human Ecology: a Problem in Synthesis". *Science*, 120(3128):959–963.

———. 1956. "Some Notes on the Ecology of Ecologists". *Scientific Monthly*, 83(1):22–27.

———. 1957. *The Ecology of Man*. Oregon State System of Higher Education, Eugene, Ore., Condon Lectures, 61 pp.

———. 1959. *Deserts on the March*. (Third edition.) Univ. Oklahoma Press, Norman, Okla., xiii + 178 pp.

Shelford, V. E. 1944. "Deciduous Forest Man and the Grassland Fauna". *Science*, 100:135–140, 160–162.

Silone, Ignazio. 1968. "Re-thinking Progress". *Encounter*, 30(3):3–12; (4):27–40.

Skinner, B. F. 1971(1972). *Beyond Freedom and Dignity*. Alfred A. Knopf, New York, 225 pp.

Snow, C. P. 1959. *The Two Cultures and the Scientific Revolution*. The Rede Lecture, 1959. Cambridge Univ. Press, London, New York, 58 pp.

Sollers, Philippe. 1972. *Lois*. Seuil, Paris, 143 pp.

Sorre, Max. 1947–48. *Les fondements de la géographie humaine*. Librairie Armand Colin, Paris, I: 447 pp.; II: 608 pp.

Stamp, L. Dudley. 1950. *The Land of Britain: Its Use and Misuse*. Longmans, Green and Co., Ltd., London, viii + 507 pp.

——— (ed.). 1961. *A Glossary of Geographical Terms*. Longmans, Green and Co., London (John Wiley & Sons, New York), xxx + 539 pp.

Strahler, Arthur N. 1965. *Introduction to Physical Geography*. John Wiley & Sons, Inc., New York, ix + 455 pp.

Sue, Eugène. 1842–43(1850). *The Mysteries of Paris* (transl. by Charles Rockford). Charles Daly, London, 533 pp.

Tansley, A. G. 1935. "The Use and Abuse of Vegetational Concepts and Terms". *Ecology*, 16:284–307.

Teilhard de Chardin, Pierre. 1955(1961). *The Phenomenon of Man*. Harper Torchbooks, New York, 318 pp.

Thomas, William L., Jr. (ed.). 1956. *Man's Role in Changing the Face of the Earth*. Univ. Chicago Press, Chicago, xxxviii + 1193 pp.

Turnbull, Colin M. 1962. *The Forest People*. Nat. Hist. Library, Anchor Books, Doubleday & Co., Inc., New York, xii + 305 pp.

UNESCO. 1969. *Conférence intergouvernementale d'experts sur les bases scientifiques de l'utilisation rationnelle et de la conservation des ressources de la biosphère*. Rapport final.

Paris, 4–13 septembre 1968. UNESCO, Paris, 38 pp. +
annexes.

————. 1970. *Interdisciplinary Symposium on Man's Role in
Changing the Environment: Architecture and Urbanism for
Growth and Change.* Otaniemi, Helsinki, June 1970.
UNESCO, Paris, Final report, 8 pp. + annexes.

————. 1972. *International Co-ordinating Council of the
Programme on Man and the Biosphere* (MAB). First ses-
sion. Final report. 9–19 November 1971. UNESCO, Paris,
65 pp.

United Nations. 1972. *Environment-Stockholm. Declaration,
Plan of Action.* Center for Economic and Social Information, U. N. European Headquarters, Geneva, 24 pp.

Van Valkenburg, S. 1950. "The World Land Use Survey".
Economic Geography, 26(1):1–5.

———— (ed.). 1952. *Report of the Commission on World
Land Use Survey for the Period 1949–1952.* Int. Geogr.
Union and UNESCO, Worcester, Mass., 23 pp.

———— (ed.). 1956. *Report of the Commission on Inventory
of World Land Use.* Int. Geogr. Union and UNESCO,
Twentieth Century Fund, New York, 67 pp.

Verhaeren, Emile. 1895(1923). "Les villes tentaculaires". In:
Oeuvres complètes (1923), Mercure de France, Paris, Volume I, pp. 95–202.

Vernadsky, W. 1945. "The Biosphere and the Noösphere".
Amer. Scientist, 33:1–12.

Vidal de la Blache, Paul. 1896. "Le principe de la géographie
générale". *Ann. de Géogr.,* 15 janvier, p. 129.

————. 1911. "Les genres de vie dans la géographie humaine". *Ann. de Géogr.,* 20:193–212; 289–304.

————. 1922(1926). *Principles of Human Geography.*
(Transl. by Millicent Todd Bingham; ed. by E. de Mar-
tonne.) Holt, New York.

Ward, Barbara. 1961. The Massey Lectures for 1961. *The
Rich Nations and the Poor Nations.* CBC, Toronto, xi + 97
pp.

117

Ward, Barbara, and René Dubos. 1972. *Only One Earth.* W. W. Norton, New York, xxv + 225 pp.

Watson, James D. 1968. *The Double Helix.* Atheneum, New York, xvi + 237 pp.

Weaver, John E. 1954. *North American Prairie.* Johnsen Publ. Co., Lincoln, Nebraska, xi + 348 pp.

Westing, A., and E. W. Pfeiffer. 1972. "The Cratering of Indochina". *The Explorer,* 14(3):8–14.

Zola, Emile. 1888(1924). *La Terre* (transl. by Ernest Dowson). Boni & Liveright, New York, xxiii + 539 pp.